D1480354

Mentoring Strategies To Facilitate the Advancement of Women Faculty

CE LIVRE A ÉTÉ LU PAR
THIS BOOK WAS READ BY

NOM/NAME   DATE

_____   _____
_____   _____
_____   _____
_____   _____
_____   _____
_____   _____

uOttawa

Centre de leadership scolaire
Centre for Academic Leadership

EX LIBRIS

Reference

Ce livre appartient à
This book belongs to

uOttawa

Centre de leadership scolaire
Centre for Academic Leadership

ACS SYMPOSIUM SERIES **1057**

Mentoring Strategies To Facilitate the Advancement of Women Faculty

Kerry K. Karukstis, Editor
Harvey Mudd College

Bridget L. Gourley, Editor
DePauw University

Miriam Rossi, Editor
Vassar College

Laura L. Wright, Editor
Furman University

**Sponsored by the
ACS Division of Chemical Education**

American Chemical Society, Washington, DC

Library of Congress Cataloging-in-Publication Data

Mentoring strategies to facilitate the advancement of women faculty / Kerry K. Karukstis ...
[et al] ; sponsored by the ACS Division of Chemical Education.
     p. cm. -- (ACS symposium series ; 1057)
  Includes bibliographical references and index.
  ISBN 978-0-8412-2592-3
  1. Women chemists--Vocational guidance. 2. Universities and colleges--Faculty.
  3. Mentoring in science. 4. Mentoring in the professions. I. Karukstis, Kerry K.
  II. American Chemical Society. Division of Chemical Education.
  QD39.5.M46 2010
  540.71'55--dc22
                                          2010048080

The paper used in this publication meets the minimum requirements of American National
Standard for Information Sciences—Permanence of Paper for Printed Library Materials,
ANSI Z39.48n1984.

Copyright © 2010 American Chemical Society

Distributed by Oxford University Press

All Rights Reserved. Reprographic copying beyond that permitted by Sections 107 or 108
of the U.S. Copyright Act is allowed for internal use only, provided that a per-chapter fee of
$40.25 plus $0.75 per page is paid to the Copyright Clearance Center, Inc., 222 Rosewood
Drive, Danvers, MA 01923, USA. Republication or reproduction for sale of pages in this
book is permitted only under license from ACS. Direct these and other permission requests
to ACS Copyright Office, Publications Division, 1155 16th Street, N.W., Washington, DC
20036.

The citation of trade names and/or names of manufacturers in this publication is not to be
construed as an endorsement or as approval by ACS of the commercial products or services
referenced herein; nor should the mere reference herein to any drawing, specification,
chemical process, or other data be regarded as a license or as a conveyance of any right
or permission to the holder, reader, or any other person or corporation, to manufacture,
reproduce, use, or sell any patented invention or copyrighted work that may in any way be
related thereto. Registered names, trademarks, etc., used in this publication, even without
specific indication thereof, are not to be considered unprotected by law.

Foreword

The ACS Symposium Series was first published in 1974 to provide a mechanism for publishing symposia quickly in book form. The purpose of the series is to publish timely, comprehensive books developed from the ACS sponsored symposia based on current scientific research. Occasionally, books are developed from symposia sponsored by other organizations when the topic is of keen interest to the chemistry audience.

Before agreeing to publish a book, the proposed table of contents is reviewed for appropriate and comprehensive coverage and for interest to the audience. Some papers may be excluded to better focus the book; others may be added to provide comprehensiveness. When appropriate, overview or introductory chapters are added. Drafts of chapters are peer-reviewed prior to final acceptance or rejection, and manuscripts are prepared in camera-ready format.

As a rule, only original research papers and original review papers are included in the volumes. Verbatim reproductions of previous published papers are not accepted.

ACS Books Department

Contents

National Initiatives

Recommendations for Individuals

Indexes

Chapter 1

A Brief Synopsis of Volume Highlights

Kerry K. Karukstis*

Professor of Chemistry, Harvey Mudd College, Claremont, CA
*Kerry_Karukstis@hmc.edu

Compelling evidence exists to support the hypothesis that both formal and informal mentoring practices that provide access to information and resources are effective in promoting career advancement, especially for women. Such associations provide opportunities to improve the status, effectiveness, and visibility of a faculty member via introductions to new colleagues, knowledge of information about the organizational system, and awareness of innovative projects and new challenges. This volume developed from the symposium "Successful Mentoring Strategies to Facilitate the Advancement of Women Faculty" held at the 239th National Meeting of the American Chemical Society in San Francisco in March 2010. The organizers of the symposium, also serving as the editors of this volume, aimed to feature an array of successful mechanisms for enhancing the leadership, visibility, and recognition of academic women scientists using various mentoring strategies. It was our goal to have contributors share creative approaches to address the challenge of broadening the participation and advancement of women in science and engineering at all career stages and from a wide range of institutional types. Inspired by the successful outcomes of our own NSF-ADVANCE project that involved the formation of horizontal peer mentoring alliances, we have assembled this collection of valuable practices and insights to both share how our horizontal mentoring strategy has impacted our professional and personal lives and to learn of other effective mechanisms for advancing women faculty.

© 2010 American Chemical Society

Initiatives at the Institutional Level

The first section of the volume features mentoring programs developed for implementation at a particular institution or a group of collaborative institutions. As a set the contributions reflect a range of campuses and describe programs aimed at a wide range of career stages. Many of the initiatives can be adopted for different settings and thus constitute a powerful "toolkit" for institutions looking for effective formal and informal mentoring schemes to target a range of challenges.

In Chapter 2, Dr. Shannon Watt, a postdoctoral chemist at the University of Michigan, argues for the development of programs and practices aimed at enabling female doctoral-level chemists to achieve their full potential and to attain their personal and professional goals. She attributes the scarcity of endeavors focused on encouraging female doctoral-level chemists to continue in the sciences after completing their training as one of the major contributors to the leak in the career pipeline of academic women chemistry faculty. As a recipient of a prestigious National Science Foundation (NSF) Discovery Corps Fellowship, Dr. Watt was required to conduct a high-impact service project that addresses national needs. In this contributed chapter, Dr. Watt describes her establishment of the Chemistry Professional Development Organization (CPDO) at the University of Michigan in 2009 to address the professional development needs of chemistry-affiliated graduate students and postdoctoral associates, particularly women and underrepresented minorities. These needs were identified through a survey of the climate experienced by graduate women across the University of Michigan; the survey administration was conducted in conjunction with an award to the University of Michigan from the NSF-ADVANCE program that aims to increase the participation of women faculty in academic STEM careers. Data from this survey showed that a majority of the graduate students and postdoctoral scientists surveyed—regardless of demographic group—desired access to training and mentoring programs that would enable them to acquire information, build networks, and develop the necessary professional and personal skills to complement their research expertise. Dr. Watt makes the case for a mentoring initiative to augment the traditional, research-centered graduate curriculum to assist in developing additional critical professional skills. Her chapter describes the initiatives of the Chemistry Professional Development Organization and provides several evaluative measures that reflect the success of this mentoring program. Dr. Watt shares her insights and strategies for establishing such a program to suit the needs and budgets of other individuals or institutions in all STEM disciplines. Adoption of such a widespread model is likely to have a significant impact on improving the retention of women in academic careers.

Chapter 3 describes a collaborative effort at two neighboring liberal arts colleges – Union College and Skidmore College – to develop mentoring networks that provide faculty with a variety of mentors who can share their successes and challenges. The chapter describes a range of formal and informal mentoring activities that offer faculty throughout the ranks with many opportunities to build a network of STEM women who can serve a variety of functions such as role models, mentors, sounding boards, and advocates. This initiative complements the pre-existing individual mentoring programs at each institution and particularly aims to provide effective information and resources about the tenure and promotion process for female assistant and associate professors. One of the merits of the collaboration is that the two institutions bring different experiences and strengths to the project as a consequence of their distinct origins. One campus was originally a women's college that traditionally emphasized the arts and humanities but now has an increased role of the science, technology, engineering, and mathematics (STEM) disciplines in its curriculum, while the other institution was a formerly all-male college that historically has had a strong natural science and engineering orientation. Such different perspectives broaden the utility of the mentoring tools developed for wider audiences.

The collaboration among three campuses of a single institution – Rutgers University – is featured in Chapter 4. To provide the context for the mentoring program showcased in this chapter, the authors first present a clear description of the unique organization of this university and the elements of each campus' chemistry department. A significant array of initiatives are enumerated that aim to drive institutional transformation that will promote the participation and advancement of women in science, engineering, and mathematics on all three campuses of Rutgers University. Of particular focus in the chapter is the RU FAIR (Rutgers University for Faculty Advancement and Institutional Re-imagination) Professorship program which enables individual faculty to take on leadership roles in advancing women's participation in the sciences. One faculty member on each of the three Rutgers' campuses is awarded the RU FAIR professorship and serves as a as university leader to foster mentoring, promote diversity, facilitate communication among geographically dispersed faculty, and mediate between faculty and administration. Such leadership can take a variety of forms, including organizing a series of professional development and leadership workshops that include sessions on leadership training, writing, grantsmanship, and faculty-to-faculty coaching (co-mentoring). Additionally, RU FAIR Professors have also encouraged research on the institutional climate for increasing women and minority faculty's participation and advancement in the sciences. While RU FAIR professors are highly visible mentors and advocates for women faculty on their campus, they authors outline some of the challenges of placing such significant responsibility for institutional transformation in a few key individuals.

This section of the volume concludes with a contribution from Auburn University featuring their NSF-funded ADVANCE project aimed at the establishment of a "small wins" approach to influence lasting change in the culture and climate of the STEM disciplines at Auburn. This chapter advocates for incremental changes with widespread and long-term impacts to eventually transform an institution. These small wins are practices implemented at the departmental, center, or college level that result in greater buy-in from all administrative levels and ultimately more substantial institution-wide transformation. Of particular interest in this chapter is a cost-benefit analysis of best practices employed at other ADVANCE-funded institutions. Using ADVANCE program websites and published materials, the most common faculty

3

development initiatives geared for women were analyzed and categorized. Several general categories were noted: implementation of mentoring practices, creation of family-friendly policies, organization of training programs aimed at raising awareness of gender bias for various campus constituencies, design of department-wide workshops that highlight the scholarship of female faculty and provide guidance on improving departmental climate; creation of departmental policies and resources that aim to improve the recruitment and retention of female faculty; and creation of funding opportunities aimed at recruitment and retention of female faculty. A cost-benefit analysis was conducted using a web-based survey instrument to identify those practices that required the fewest resources and contributed the most to the improvement of the university climate and community. Of the 29 initiatives evaluated, mentoring programs represented over half of the most highly ranked practices employed at other universities. The chapter describes how this information was used to develop and implement effective programmatic changes at Auburn University.

Multi-Institutional and Interinstitutional Initiatives

The second section of the volume describes mentoring activities at various collections of similar institutions. Chapter 6 examines faculty mentoring at two-year institutions, the segment of the higher education system that represents 34% of the nation's post-secondary institutions and serves a substantial portion of the undergraduates in the United States. With over 1200 institutions, two-year colleges exhibit a diversity of sizes, locations, and program offerings to meet the needs of the regions they serve. Given the extensive array of two-year campuses, a variety of faculty mentoring approaches is anticipated. To get a flavor of how wide-ranging such faculty development efforts might be, the author of this chapter asked eight female chemistry faculty members at different two-year colleges to share their perspectives on the status of women faculty on their campuses. While the situation does indeed vary from campus to campus, this collection of women faculty generally report strong satisfaction in their careers, in the faculty development expectations and offerings on their campuses, and in the institutional mentoring programs available for new faculty. The combination of institutional mission, high numbers of female faculty members even at all levels, and the range of internal and external professional development opportunities suggest a supportive climate that enables two-year college female faculty to prosper. The scope of formal and informal mentoring initiatives present at the campus level and in conjunction with professional societies is highlighted.

Chapters 7 through 9 represent contributions from women full professors in chemistry and physics at liberal arts colleges. All of the authors were participants in the NSF-ADVANCE funded project that is described in Chapter 10, and each recognized the importance of adding to the knowledge base of mentoring strategies and career development resources that contribute to the advancement of academic women at liberal arts institutions.

Chapter 7 describes institutional and departmental mechanisms which support women faculty in chemistry at liberal arts colleges at all stages of their career, from the pretenure years through retirement. As the authors are all senior women in chemistry, they particularly focus on specific recommendations of policies aimed at supporting women at this career stage. One of their major themes is the need for flexibility in granting resources, developing policies, and providing infrastructure for the professional development of women. The authors also suggest a variety of ways that women can and do support one another. Given the employment experience of the authors, they discuss how shared/split academic positions can enable more academic women to enjoy a better work-life balance and offer insights to both advantages and shortcomings of these hiring arrangements.

In Chapter 8, five accomplished senior female physics faculty describe the unique challenges and demands of senior women scientists at liberal arts institutions. They particularly cite the ways in which their horizontal mentoring alliance helped each participant to successfully navigate a variety of professional and personal issues. One of the areas of professional concern for this group of women was maintaining their research vitality over the course of an entire career. Each pondered next steps such as whether to continue to extend current work with the goal of remaining on the cutting edge of the field, change to a new sub-field to explore new areas of interest, or even to shift gears to pursue less traditional research in pedagogical arenas. The chapter highlights the way in which the alliance was instrumental in strengthening each member's professional research by the answering the question of "What next?" in different ways. Through the experiences of the alliance members, this paper makes a strong case for sustaining and propagating similar networks and suggests some initial steps to achieve this continuity without the need for significant external funding.

Indeed, Professor Carol Ann Miderski explores in Chapter 8 one such mechanism for continuing and expanding the practice of horizontal peer mentoring across a number of institutions within a close geographical region. Professor Miderski's home institution is situated in a region with a significant number of small undergraduate-focused campuses with similar low numbers of female faculty in chemistry. To overcome such professional isolation and continue the benefits that she has experienced in her horizontal network, Professor Miderski described her initiation of The Women Chemists Web in 2009 to bring women faculty from regional colleges together to get to know each other and to develop a resource network. She designed the group with the objective of serving as a source of outside perspectives, fresh ideas, and alternative strategies for facing the academic, professional and personal challenges encountered in small college environments. This chapter shares some of the insights gained by exploring the most commonly-cited vexing issues for women faculty and offers some of the mechanisms by which The Women Chemists Web will serve as a resource for participants.

National Initiatives

The third section of the volume showcases two mentoring initiatives administered at the national level for women in academe. In Chapter 10 the editors of this volume describe their project funded by the National Science Foundation

ADVANCE Partnerships for Adaptation, Implementation, and Dissemination (PAID) program to test a horizontal mentoring strategy for senior women faculty in chemistry and physics at liberal arts colleges. The project, Collaborative Research for Horizontal Mentoring Alliances, focuses on the distinctive environments of undergraduate liberal arts institutions and the challenges faced by senior women faculty on these campuses to attain leadership roles and professional recognition. Four five-member alliances of senior women faculty members at different institutions were formed for the purpose of "horizontal mentoring" to enhance the leadership, visibility, and recognition of participating faculty members. The chapter describes the rationale for the horizontal mentoring approach and the key elements of the alliance structure to insure the effectiveness of this form of peer mentoring. The chapter also describes the mechanics of alliance formation, the professional development activities of alliance gatherings, and the professional and personal benefits of participation cited by the twenty women faculty involved in the project. The benefits include the added confidence to seek leadership positions, enhanced visibility and recognition on campus, encouragement to seek and accept external recognition, and support to pursue new directions. The authors of this chapter, as the editors of this book, have found the horizontal mentoring project to be one of the most powerful undertakings of their professional careers.

Chapter 11 highlights the decade-long faculty development efforts of COACh, the Committee on the Advancement of Women Chemists. COACh is an organization that focuses on developing and implementing programs to increase the career success of women chemists in academia.

Included among the many activities sponsored by COACh are workshops that provide negotiation, management, and leadership skills to help women achieve their professional goals as faculty in the chemical sciences. These workshops are a form of group mentoring where a protégé has access to a group of experienced individuals working together to provide career information to the protégé with each mentor contributing her unique talents to the group. The chapter examines women chemists' mentorship experiences by drawing from information gained from surveys and interviews of individuals who participated in the COACh workshops over the past decade. The authors share their insights on a variety of aspects of mentoring, including the effectiveness of formal mentoring programs, the changing mentor/mentee role over the course of a career, why mentoring often doesn't happen, and what factors can contribute to having a positive mentoring experience. The particular ways that COACh has promoted mentoring and the outcomes of such efforts are also discussed. The authors conclude their work by indicating the mentoring research that still needs to be completed and sharing lessons for policy and action.

Recommendations for Individuals

In the final section of the volume we address two of the key professional challenges that academic women routinely find vexing – integrating work and a personal life and enhancing one's professional presence. These topics transcend institutional type and even career stage. In Chapter 12, Drs. Millard and Mills advocate for the importance of faculty well-being to maintain both professional productivity as well as personal satisfaction and to cope with both time and stress management. The chapter begins with a consideration of the hidden consequences of failing to achieve an acceptable balance of professional and personal commitments and a discussion of the practice of "bias avoidance" that leads to behavior that minimizes or hides the impact of family life on academic commitments. The particular challenges faced by senior women and those in the sciences are further outlined. The bulk of the chapter provides a wealth of useful tips for better integrating one's personal and professional lives. Many of the suggestions are derived from the personal experiences of the authors and offer successful strategies for simultaneously achieving fulfillment in one's career as well as contentment in one's personal life. The very useful exercise of determining one's "chaos coefficient" is an effective first step toward achieving balance. As the authors note, incorporating personal needs into the equation is essential for attaining the most sustainable lifestyle.

In Chapter 13, Dr. Millard continues to provide insights gained from her professional career as she offers suggestions for enhancing one's professional impact and acquiring the leadership positions and recognition commensurate with one's expertise. While women in science fields anticipate being judged on their professional credentials, Dr. Millard reminds us that other unexpected factors may be used in assessing our professional competence. For example, students and colleagues may use our physical appearance, body language and nonverbal cues, and attire to judge our professional capabilities. In today's electronic world where impressions are made in the absence of face-to-face interactions, Dr. Millard makes a strong case for maintaining a strong virtual presence. The art of effective self-promotion - communicating one's strengths and accomplishments to others in a sincere way without appearing to be bragging - is also a skill that women faculty should master. While it's wonderful when others notice another's achievements, individual faculty are in the best position to share their accomplishments with others. The chapter concludes with some expert advice for those faculty members privileged to be in leadership positions, namely understanding the responsibilities associated with holding prominent roles on campus.

Final Thoughts

We sincerely thank all of the contributors to this volume. This compendium of successful mentoring practices to enhance the leadership, visibility, and recognition of academic women in science and engineering emphasizes the importance of the collective efforts of the academic community to broaden the participation and advancement of women faculty. It is our sincere hope that readers of this volume will find valuable information that assists individual faculty members in their careers and inspires institutions to provide the resources that enable every faculty member to flourish. An investment by an institution in the continuous development of a faculty member's career will have a broad impact not only on the individual faculty member, but also on his or her colleagues and students and on the ability of the institution to attract and retain excellent faculty and students.

This material is based upon work supported by the National Science Foundation under Grants No. NSF-HRD-0618940, 0619027, 0619052, and 0619150. Any opinions, findings and conclusions or recommendations expressed in this material are those of the author(s) and do not necessarily reflect the views of the National Science Foundation (NSF).

Initiatives at the Institutional Level

Chapter 2

Facilitating the Advancement of the Next Generation of Women Faculty: Female Graduate Students and Postdoctoral Associates

Shannon Watt*

Department of Chemistry, University of Michigan, 930 N. University Ave.,
Ann Arbor, MI 48109-1055
*email: shwatt@umich.edu

It is critical to facilitate the advancement of female faculty by developing and promoting successful mentoring strategies at all educational and professional levels. This is especially true during doctoral and postdoctoral study, when potential future faculty members evaluate their career options and academic environments on a daily basis. A recently-established program at the University of Michigan focuses on addressing the identified professional needs of chemistry graduate students and postdoctoral associates, particularly women and underrepresented minorities, in areas not commonly addressed during doctoral and postdoctoral training. This chapter discusses the establishment of the student- and postdoc-led Chemistry Professional Development Organization, which has developed a variety of data-driven programs to equip chemists of all backgrounds—especially those in underrepresented groups—with the tools to reach their personal and professional goals, including pursuit of faculty careers.

Background

Although women currently earn 50% and 36% of bachelor's and doctoral degrees in chemistry (1), they remain underrepresented at almost all levels of academic faculty and administration, as well as in industry and government (2). Table I shows the representation of women among chemistry faculty members based on the highest degree awarded by their institution (2, 3). While women

© 2010 American Chemical Society

of all races-ethnicities are underrepresented compared to their proportion of the U.S. population, this is particularly true for women of color: African-American, Hispanic/Latina, and Native American females comprised *less than 1%* (22 of 2787) of tenure-track faculty at the top 100 chemistry departments (by National Science Foundation research expenditures) in 2007 (*4*).

The dearth of females in science, technology, engineering, and mathematics (STEM) fields, including chemistry, has often been attributed to few women pursuing careers in technical fields or to insufficient 'lag time' to allow the employment rates to 'catch up' to the educational rates. At least in chemistry, however, these suggestions are not supported by statistics. Women have earned at least 25% of all Ph.D.s in chemistry since the late 1980s—rising above 30% by the late 1990s—yet after three to four tenure cycles comprise only 12% to 18% of full professors at institutions granting at least a baccalaureate degree (*1–3*). In fact, the increase in representation of women among chemistry doctoral degree holders in recent decades is itself potentially misleading because it is due not only to the higher number of female graduates (723 in 2005 vs. 362 in 1985) but also to the decreasing number of male graduates (1,403 in 2005 vs. 1,474 in 1985) (*5*).

The underrepresentation of women in the ranks of academic chemistry faculty can be attributed in large part to significant leaks in the career pipeline that represents progression from (pre)undergraduate training through the professional ranks. Leaks occur before, during, and immediately following the Ph.D. as a result of self-selection of qualified individuals out of these careers (*6, 7*). For example, despite earning 30-32% of chemistry doctorates during the late 1990s (*1*), women comprised just 22% of chemistry postdocs in 2002 (*8*) and 18% of applicants for research-intensive chemistry faculty positions from 1999-2003 (*9*). Reasons for this exodus from academic careers are likely to be as numerous and complex as the women making the decisions. In recent years, a number of studies have sought to describe underlying themes related to the experiences of graduate students (and occasionally postdocs) in all fields (*10*), in the sciences (*11*), and in chemistry (*8, 12–14*). Work undertaken in conjunction with this project also evaluates demographic differences in key experiences, values, and factors that impact the career choices of graduate students and postdocs within a large U.S. chemistry department (*15*). Ongoing research in this area is critical for developing data-driven programs and policies that increase retention of qualified doctoral-level chemists from all backgrounds.

Why is it necessary to broaden the participation of chemists in underrepresented groups? Research has demonstrated that groups comprised of individuals with a variety of perspectives outperform those comprised of like-minded thinkers in terms of problem-solving and innovation (*16, 17*). Among other factors, an individual's gender, race-ethnicity, place of origin, and socioeconomic status contribute to his or her individual approach to problem-solving. Thus, it stands to reason that the chemical enterprise best positions itself for success in innovation and problem-solving when it includes contributions from individuals with diverse experiences and backgrounds.

Successful programs exist to promote and support women in science from preschool through the workplace, but they are rare at the graduate and postdoctoral levels. The scarcity of endeavors focused on encouraging female doctoral-level

Table I. Representation of Women at Various Chemistry Faculty Ranks as a Function of Institution Type (2, 3)

	Assistant Professor	Associate Professor	Full Professor
Bachelor's-granting[a]	37%	36%	18%
Master's-granting[a]	33%	31%	16%
Doctorate-granting[b]	27%	23%	12%

[a] Data from 2005.   [b] Data from 2009.

chemists, particularly women of color, to continue in the sciences after completing their training is a major contributor to the leak in the aforementioned pipeline to both academic and non-academic careers. These students already have demonstrated an interest in science careers, yet they face imminent decisions about continuing to pursue these paths. Consequently, it is imperative that programs are created to repair this leak, retain chemists who obtain advanced degrees, and establish underrepresented groups in scientific careers. Such endeavors must focus on developing programs and practices that engage, enable, and inspire female and minority doctoral-level chemists to achieve their full potential and to attain their personal and professional goals. In the long term, policies and practices that broaden the participation of women will strengthen the domestic technical workforce and contribute to a level playing field for *all* chemists.

Mentoring plays a key role in career success (*8*). In an ideal world, all graduate students and postdocs would enjoy ample and uniform access to training and mentoring experiences that help them acquire information, build networks, and develop all of the necessary professional and personal skills to complement their research prowess. However, such aspects of training and mentoring often are not included in the traditional, research-centered curriculum. Recent data show that a majority of the graduate students and postdocs surveyed—regardless of demographic group—desired access to such complementary programs (*15*). In addition, a significant percentage of these students and postdocs, particularly women, lack mentors (*15*). Anecdotally, it appears that these trainees may not recognize the need for mentoring or may be unsure how to identify and establish a relationship with appropriate mentors. If, for whatever reason, mentoring is not readily available or accessible, how can students and postdocs fill the gap for themselves?

Complementary Initiatives

Across the United States

A number of initiatives have focused on broadening women's participation at the graduate and/or postdoctoral levels. The Committee on the Advancement of Women Chemists (COACh) has expanded its repertoire of professional skills workshops for female faculty to include sessions specifically designed to help female and underrepresented graduate students and postdocs build their negotiation skills and develop strategies for career success. These workshops are offered at professional society meetings (including those of the American Chemical Society) and in conjunction with individual organizations or institutions. MentorNet pairs female and underrepresented minority graduate students, postdocs, and early-career faculty at partner institutions with senior mentors in one-on-one 'e-mentoring' relationships. A number of campus-based programs—for example, the Stanford Chemistry Women's Committee on Graduate Life, the Georgia Tech Women in Chemistry Committee, and several Association for Women in Science (AWIS) and Iota Sigma Pi chapters—have also taken steps to improve various aspects of female graduate students' and/or postdocs' experiences. Such initiatives include support programs, career-related events, issue-focused discussions, and maternity or parental leave policies or guidelines.

At the University of Michigan

Since its founding in 2001 as part of a larger National Science Foundation effort to increase the participation of women faculty in academic STEM careers, the University of Michigan (UM) ADVANCE program has developed a number of successful department-, university-, and nationally-based initiatives to effect institutional climate change. This work, including a survey of the climate experienced by graduate women across UM, has set the stage for the development of new, complementary endeavors to increase the participation of graduate students and postdocs from underrepresented groups.

The UM chemistry department has also implemented effective programs over the past several years to increase the gender and racial-ethnic diversity of its faculty and to enhance the departmental climate for diverse populations. As a result of a UM ADVANCE-sponsored Departmental Transformation Grant, the faculty hiring process was redesigned to foster diversity; departmental policies were modified to be more democratic and transparent; and mentoring efforts and departmental climate were enhanced to foster the success of junior faculty, especially women. This program resulted in a substantial increase in the representation of women among department faculty, from 2.5 female professors in 2001 to 8 (of 37) in 2009. It also set the stage for the establishment of a similar program focused on graduate students and postdocs.

The University of Michigan Chemistry Professional Development Organization

The project described herein extends the UM ADVANCE and chemistry faculty development initiatives described above to meet the needs of chemistry graduate students and postdocs. The department is home to 76 postdocs, including 21 women, 3 underrepresented minorities, and 47 foreign nationals. Of 193 graduate students enrolled in chemistry, 107 are women, 10 are underrepresented minorities, and 55 are foreign nationals. In addition to those enrolled in the chemistry department, 76 graduate students are enrolled in other UM departments (e.g., Macromolecular Science and Engineering, Applied Physics, and Biological Chemistry) but have research advisors with primary appointments in chemistry. A number of these 'chemistry-affiliated' students choose to take part in project-related activities.

Origin and Establishment

The Chemistry Professional Development Organization (CPDO) at the University of Michigan was founded in 2009 to address the professional development needs of chemistry graduate students and postdoctoral associates, particularly women and underrepresented minorities; these needs were identified through a department-wide assessment and series of listening sessions. The organization is one component of a National Science Foundation Discovery Corps Fellowship project to evaluate and address a number of factors related to the experiences of members of underrepresented groups within the UM chemistry department. The CPDO's programs and activities are meant to complement more traditional mentoring strategies rather than to replace them entirely. This program is one of several potential tools in a graduate student's or postdoc's mentoring toolbox.

Before constituting the organization, all graduate students and postdoctoral associates affiliated with the department (both chemistry enrolled/appointed and 'chemistry-affiliated') were asked to participate in a confidential, anonymous online assessment. Disaggregated data from this study were used to evaluate the personal and professional needs and goals of the participants and the ability of current standards and practices to enable individuals to reach their goals. The data clearly indicate that women often have significantly different support and professional development needs from their male counterparts, including different levels of expressed interest in co-curricular programs (18). A subsequent series of four listening sessions was held to affirm the assessment findings; refine program goals; promote active engagement by and support from graduate students, postdocs, faculty, staff, and administrators; and recruit members for the organization. Seven founding graduate students and postdocs established the group and named it the Chemistry Professional Development Organization.

Organizational Structure

UM chemistry and 'chemistry-affiliated' graduate students and postdocs of all backgrounds are welcome to join the organization. CPDO membership is not required to participate in our activities. We strive to include as many people as possible from diverse backgrounds (i.e., gender, citizenship, race-ethnicity, seniority, sub-disciplines, and departments of enrollment/appointment). Current members include 10 women, 2 underrepresented minorities, and 5 foreign nationals. Twice each year, new members are recruited to serve renewable one year terms. These staggered terms allow for continuity, new member training, knowledge transfer, and development of leaders from within. The addition of a second membership cohort in early 2010 increased the group size from 7 to 13; as of this writing, members of a third cohort are beginning their terms.

The CPDO has chosen to adopt a relatively flat structure, with the only official role being that of the organization's chair; however, the group's setup allows for the installation of co-chairs or more traditional officers as it evolves. Members take turns arranging and chairing the hour-long biweekly organization meetings, which distributes responsibility throughout the group and allows each member to develop his/her leadership skills in a low-pressure setting. Initially, the size and interests of the group led each member to take responsibility for a particular area (seminar series, website, networking events, etc.). Now that the group is larger and more established, this is no longer necessary. Each member commits to organizing at least one event per year, and more senior members mentor newcomers as they begin to plan events with the aid of CPDO-developed guides, checklists, and templates. In addition, members have access to an internal resource website containing a digital archive of all CPDO records (recruiting materials, event details, meeting minutes, etc.). These resources empower new members to take ownership of their event while minimizing the time investment and potential intimidation associated with a new undertaking. A number of the resources developed in conjunction with this project may be made available to leaders of similar programs upon request.

Thematic Initiatives

Based on assessment data and information from listening sessions, we chose to establish three main focus areas: career exploration; professional skill development; and community-, communication-, and resource-building. To date, we have hosted 22 events and led several projects within one or more of these broad areas. Speakers are identified within the department, on campus, in the region, or across the country by word-of-mouth, through web research, or via networking. Speaker travel and event-related expenses (e.g., refreshments) are funded by the Discovery Corps grant. CPDO-sponsored programs are open to all chemistry graduate students and postdocs, regardless of gender or race-ethnicity; in fact, many of our participants are Caucasian men.

Career Exploration

In the course of their training, students and postdocs often witness aspects of their faculty advisors' professional—and sometimes personal—lives. At least among female graduate students, however, recent research suggests the presence of a disconnect between students' perceptions of different careers and the experiences of women in those careers, especially in academia (*19*). Based on their daily observation of faculty life at a research-intensive university, students may focus more on the challenges of a tenure-track research career (long hours, grant deadlines, increased competition for funding) rather than identifying with potential rewards and benefits (flexible schedules, academic freedom, and supervising students in the lab). In addition, standard curricula often do not afford the opportunity to explore career options outside the research-focused academic tenure stream. As the number of chemistry doctoral degree holders far outstrips the number of tenure-track positions available (an imbalance that may only increase in the future), it is critical that chemists of all backgrounds be able to make informed career decisions.

To that end, many of our events focus on career exploration. We have organized speakers or panel discussions on careers at teaching-focused academic institutions, in patent law, in federal research labs, in science policy, and in industry. At these sessions, chemical professionals share their career paths; discuss strategies for success and for navigating challenges such as work-life balance; and answer participants' questions. We have also hosted five external visitors (four from academia, one from a federal lab) as part of the CPDO Seminar Series. In addition to meeting with UM faculty members and presenting their research in a department seminar, each speaker spends one to two days meeting with students and postdocs over meals; participates in a networking reception; and gives presentations on topics complementary to his or her research. Examples of such presentations include:

- diversity in science,
- women in academia,
- creating innovative undergraduate courses at a research university,
- developing a career in the U.S. from an international perspective,
- comparisons between academic and industry careers,
- comparisons between government research and science policy careers,
- the role of service in a chemist's career, and
- several (different) perspectives on work-life balance.

All of these sessions allow participants to more fully explore their own places in the chemical enterprise, whether through academia or another career path. The events, which often include individual or small group meetings, also afford participants an opportunity to make connections with potential mentors outside UM.

Professional Skill Development

Chemistry doctoral programs excel at training students and postdocs to become highly skilled and independent researchers. Many trainees also have an opportunity to build skills complementary to their research; for example, they may serve as substitute lecturers, assist their advisors with grant-writing and manuscript review, or mentor junior students in the lab. However, these opportunities tend to be specific to the individuals and situations involved. Therefore, we offer programs that allow all participants to build the so-called "soft skills" (e.g., communication, self-differentiation, effective teamwork, etc.) that are becoming increasingly critical for career success.

A series of events has focused on the academic job search from a variety of perspectives. At one such event, a panel of UM chemistry faculty members—a recently hired junior professor, a search committee chair, and a department chair—discussed the mechanics of applying for tenure-track research faculty positions. A follow-up event featured two postdocs, who simulated the research proposal portion of a faculty candidate interview by giving mock presentations to an audience that included several professors volunteering as 'search committee' members. These faculty members helped the audience to understand a heretofore mysterious process by posing questions commonly asked in such sessions, providing feedback on the proposals and presentations, and sharing strategies from their experiences. Soon-to-depart chemistry postdocs who have successfully obtained a variety of administrative, teaching-focused, and research-focused academic appointments have served on annual panels to discuss the details of their recent job search and share tips for success. Postdocs on the most recent panel also shared their application materials to illustrate appropriate approaches to applying for various kinds of positions.

Two other skill-building events have leveraged innovative programs to teach more generally-applicable skills. In collaboration with several campus entities, we hosted two facilitators who present workshops through the Committee on the Advancement of Women Chemists. These facilitators presented a half-day negotiation skills workshop that allowed participants to explore key elements of negotiation, assess their own conflict resolution styles, develop strategies for approaching negotiations, and practice via case studies. We also collaborated with the UM Center for Research on Learning and Teaching (CRLT) Players, a theater troupe often invited to perform their interactive skits on educational and diversity topics at national conferences and workshops. We worked with the CRLT Players to customize four such skits to illustrate the challenges that chemistry students and postdocs often face in communicating with their research advisors and lab colleagues. Coupled with written conflict resolution materials, these skits served to catalyze a guided discussion about the various dynamics involved in and strategies for successfully resolving such situations.

Community-, Communication-, and Resource-Building

It is critical that all department members enjoy equal access to communities, methods of communication, and avenues for resource dissemination. This is particularly true for postdocs, who almost always enter departments as individuals rather than in a cohort. Consequently, they lack opportunities to build relationships with faculty and colleagues through coursework, curricular requirements, or other student-focused pathways. The CPDO events and programs described above also often include one or more components of our community-, communication-, and resource-building objective. These networking-based efforts complement the department Graduate Student Council's periodic social events and the annual department-wide research symposium.

As is common in large departments, some individuals may lack opportunities to interact with peers or potential mentors. We have initiated a periodic series of networking events to increase informal interaction among graduate students, postdocs, and faculty. The events have been held at various times of day (breakfast, lunch, mid-afternoon) to facilitate participation by those with constrained schedules. We have also co-sponsored a gathering specifically for postdocs to enhance community within this population.

In addition to programming, the CPDO has collaborated with chemistry administrators to enhance various aspects of the department experience and to facilitate equal access to resources. Recent initiatives have included implementing a preparation process to facilitate the success of those participating in on-campus interviews for industrial positions and establishing online graduate student and postdoc personnel directories to foster communication. CPDO leaders disseminate career-related announcements through our email list. In most cases, podcasts of and documents from our events are posted on an internal website, enabling UM chemistry students and postdocs to utilize these resources on an ongoing basis. We also have established a CPDO website (http://www.umich.edu/~chempdo) that includes an ever-expanding list of department, campus, and external resources of relevance to chemistry students and postdocs.

Evaluation and Measures of Success

We evaluate our individual events and our overall strategic plan on an ongoing basis. Following each program, participants are asked to complete an anonymous, online evaluation that combines Likert-scale and open-ended questions about the event's interest, timing, speaker(s), format, suitability to address a previously unmet professional development need, and most and least positive aspects. Attendees are also invited to suggest topics for future events and to leave any other comments. If the commenter chooses to provide an email address, we reply to questions or suggestions by return email.

In addition to gathering formal program data, we also host informal biannual open house sessions; the purpose of these casual, drop-by events is to provide a mechanism for informal feedback, to disseminate information about the CPDO, and to recruit new members. We also debrief each event at a subsequent CPDO meeting, discussing what was successful and what might be improved upon in the future. The CPDO holds one strategic planning meeting each semester to ensure the continued relevance of our focus areas and to lay the groundwork for a cohesive series of upcoming events.

As of June 2010, 207 individual chemistry and 'chemistry-affiliated' graduate students and postdocs have participated in at least one of our 22 events, for a grand total of 703 participant-occurrences. This is only one of several possible measures of success, but it is supported by our post-event evaluation data. In aggregate, over 83% of participants who responded to evaluation surveys agree or strongly agree that an individual event addressed a previously unmet professional development need. This figure rises to 89% for career-exploration or skill-building events, indicating that these events are of particular relevance. Overall, 93% of evaluation respondents agree or strongly agree that they would recommend the events to others. Word of the organization's activities has spread, and a number of trainees from other UM departments have asked to participate in our events.

Post-event evaluations also reveal participant enthusiasm in the form of open-ended comments about the programs attended. The quotes below illustrate examples of participants' responses regarding the effectiveness of various programs.

- Regarding a panel discussion featuring faculty members from teaching-focused institutions: "[T]hey were able to provide insights regarding differences between small and large universities that we...would not have generally known to ask about."
- Regarding a career exploration event with a program officer from the National Academies: "The speaker [discussed] a job opportunity I had never thought of before. I had no idea jobs like [hers] really existed."
- Regarding a CPDO Seminar Series speaker from a research-focused university: "[I]t was exciting to hear a faculty member very happy with her position and so open about how she got there."
- Regarding a speaker on careers in federal research labs: "[She] gave a great overview of job opportunities and working environments in...national labs. I really appreciated her candor in talking about problems women have faced within those labs and ways they dealt with them to advance their careers."
- Regarding a CPDO Seminar Series speaker from a research-focused university: "[The most positive aspect was] the less formal discussion of cliques/clubs and the background politics of a faculty position."
- Regarding the panel discussion on the mechanics of applying for tenure-track research faculty positions: "Discussion of nego[t]iating a start-up package was information I'd never heard before."
- Regarding the mock faculty candidate research proposal presentations and discussion: "[It] demystified the whole idea of the job proposal talk."
- Regarding the 2010 panel with postdocs departing for academic positions: "It was helpful to realize that where you are targeting your job application to influences the process so much. The variety in the panelists helped greatly to understand the process as a whole. I'm sure...the packet [of panelists' application materials] will come in helpful too."
- Regarding the negotiation skills workshop: "[I]t helped me realize how important negotiation is, and showed me ways to be a better/effective negotiator."
- Regarding the CRLT Players skits and discussion on conflict resolution: "Open discussion about each topic was very effective as everyone can participate and give an input on the matters face[d] everyday."
- Regarding a networking event where participants were seated according to the color of their food plates: "Matching your food plate color to a particular table...enables meeting with new colleagues rather than choosing a table with your friends."

Implications for Other Departments

There is an increasing interest in broadening the participation of underrepresented groups in chemistry by recruiting and retaining diverse populations of graduate students and postdocs. This interest was reflected at the Spring 2010 Graduate and Postdoctoral Diversity Programs Summit, which fostered connections between CPDO leaders and those in similar roles at other institutions, including Stanford, Georgia Tech, the University of California San Diego, and the University of California Santa Barbara. Summit delegates agreed that these department-based programs play a significant role in enhancing women's graduate and postdoctoral experiences, increasing retention of women in chemistry careers, and working to level the playing field for all chemists. Even in economically challenging times, such programs constitute a minute fraction of a research-intensive department's budget. The costs are far outweighed by the benefits in recruitment, retention, and reputation spread by alumni as they move through the professional world.

The following are insights gained by CPDO members during the process of establishing the organization. They may be beneficial for individuals or institutions considering creating such a program, whether in chemistry or in another STEM department.

Funding

CPDO has had the luxury of ample funding as a result of the Discovery Corps grant, and matching funds from campus entities will continue to support the program into the future. These funds allow for the purchase of supplies and refreshments for events and for reimbursing speakers for travel costs. While this no doubt contributes to the organization's success, it is not necessarily a prerequisite for other programs. The availability of refreshments may easily be varied to accommodate a range of budgets: organizations can choose to host potluck or brown bag events, provide coffee and cookies, or serve full, catered meals. Similarly, organizations with limited budgets can tap into the expertise of their own senior graduate students, postdocs, and faculty members; invite speakers from other campus units or local industry; or share costs with their local American Chemical Society section or another nearby science department.

Personnel

CPDO's founding was overseen by a postdoc appointed to spend a significant fraction of her time on the project. This allowed the organization to immediately implement a high level of activity in line with the observed demand. However, similar initiatives (such as those at Stanford and Georgia Tech) have been successfully established in situations with dedicated student or postdoc leaders, strong faculty and/or staff mentoring, and a clear message from administrators regarding the project's value to the department.

CPDO receives staff support in the form of financial management and assistance with arranging travel, processing reimbursements, ordering food for events, and publicizing seminars. Again, this level of support is valuable but not absolutely necessary, especially for organizations with minimal logistical needs. However, we believe that some amount of staff support is critical for an organization run by graduate students and postdocs, who balance this service to their colleagues and department with their own research, teaching, and other duties. As such, they should be afforded sufficient resources to fulfill their mission without requiring undue sacrifice in terms of their research productivity and outside obligations. Strong communication between the organization, the staff, and the administration is key to determining what support is needed and how it will be implemented.

Communication

Both CPDO members and event participants have competing demands on their time, so communication within the group and with our constituents is key to promoting active engagement and maintaining support. Our recruitment materials and open house sessions discuss the benefits of CPDO membership, including opportunities to develop leadership experience in low-pressure settings; to build one's CV or resume; to establish or expand a professional network; and to bring to fruition events or programs of particular interest to the organizer. Program announcements explicitly describe what participants will gain (e.g., opportunities to network, to learn about a particular career path, or to acquire a new skill) by attending an event. We also ensure that all members of the department are updated regarding our activities by distributing newsletters each semester that summarize our recent events and highlight upcoming programs.

Our communication strategies appear to be effective, as evidenced by sustained high rates of participation and positive feedback. Nevertheless, communicating with a large group of people in a media-saturated society is an ongoing challenge. Based on the resources available to us, we communicate primarily through our department-wide email list and website; we occasionally invest time and resources into posting fliers for key events. We have received feedback that some graduate students, in particular, feel that email is over-utilized by various campus entities; this seems to pose a challenge in terms of connecting with the small number of department members who prefer not to communicate by email. We continue to evaluate the possibility of utilizing complementary forms of social networking, including Facebook and Twitter. Our dissemination strategies are likely to be an ongoing source of discussion, especially as forms of communication continue to evolve.

Establishing a Sustainable Organization

Setting up a new organization is both time- and energy-intensive, especially in terms of making key decisions and establishing relationships and protocols for logistical support. Several of the CPDO's founding postdocs had been involved in student-run organizations as graduate students and had experienced the fluctuations in activity and momentum inherent in any long-term initiative. To the extent possible, we were able to minimize these oscillations by deliberately establishing a simultaneously regimented and flexible organizational structure that is likely to ensure sustainability. Staggering membership terms allows for leadership continuity and transfer of institutional memory. We avoid the need to continuously 'reinvent the wheel' by maintaining templates, 'how to' documents, and a well-organized digital archive for related correspondence as we work out the logistics of programming (e.g., inviting speakers, ordering food, running meetings). While this documentation process required an up-front investment of time and energy, we've found that it allows for less-intensive upkeep in the long term and reduces the learning curve for those new to leading such programs. We also take advantage of time-saving electronic resources such as www.surveymonkey.com, www.doodle.com, and UM's web-based course and project management utility to conduct event registrations and evaluations, schedule meetings, and archive CPDO-related materials, respectively.

Establishing a data-driven vision focused on three broad areas serves as an overall strategic planning guide yet also allows us to customize programs and initiatives as new data and interests arise. Mechanisms for ongoing feedback—including post-event evaluations, internal event debriefings, and periodic strategic planning meetings—provide opportunities to evaluate events in real time while planning for the future.

Summary

To increase the representation of women in chemistry, particularly in academia, it is critical to examine and address sources of self-selection at all educational and professional levels, including graduate and postdoctoral study. Graduate students and postdocs of all backgrounds should have access to a suite of mentoring options during a time where they continually evaluate the implications of potential career choices. The student- and postdoc- led University of Michigan Chemistry Professional Development Organization complements more traditional training programs and mentoring techniques by providing opportunities to explore career options; build skills complementary to research; and leverage community, communication, and other resources. Data from a department-wide assessment, a series of listening sessions, and ongoing formal and informal evaluations inform the group's activities within these three strategic focus areas. Ample funding, staff support, clear communication, and a sustainable organizational structure allow CPDO to offer relevant and high-quality programming, enjoy robust levels of participation, and earn overwhelmingly positive feedback. Insights and strategies gained from the process of establishing such a program can be readily adapted to suit the needs and budgets of other individuals or institutions in all STEM disciplines. The widespread establishment of such programs is likely to increase the recruitment and retention of female graduate students and postdocs, equip them with the tools to reach their personal and professional goals, and level the playing field for scientists and engineers of *all* backgrounds.

Acknowledgments

This project was funded by the National Science Foundation's Discovery Corps Postdoctoral Fellowship program (award CHE-0725242), the University of Michigan chemistry department, and the University of Michigan ADVANCE and Women in Science and Engineering programs. The author wishes to thank all current and former members of the Chemistry Professional Development Organization for investing their time, energy, and talents to provide opportunities for their colleagues; Professors Abigail Stewart and Carol Fierke for their advice and mentoring during the project; and the UM chemistry staff for their support.

References

1. Hanson, D. J. Gains in chemistry grads persist. *Chem. Eng. News* **2009**, *87* (47), 38–48.
2. Heylin, M. *Chem Census 2005*; American Chemical Society: Washington, DC, 2005.
3. Raber, L. R. Women now 17% of chemistry faculty. *Chem. Eng. News* **2010**, *88* (9), 42–43.
4. Nelson, D. J.; Brammer, C. N.; Rhoads, H. *A National Analysis of Minorities in Science and Engineering Faculties at Research Universities*; Norman, OK, October 2007.
5. Heylin, M. Radical changes for U.S. science. *Chem. Eng. News* **2008**, *86* (10), 67–71.
6. Marasco, C. Numbers of women nudge up slightly. *Chem. Eng. News* **2003**, *81* (43), 58–59.
7. Handelsman, J.; Cantor, N.; Carnes, M.; Denton, D.; Fine, E.; Grosz, B.; Hinshaw, V.; Marrett, C.; Rosser, S.; Shalala, D.; Sheridan, J. More women in science. *Science* **2005**, *309*, 1190–1191.
8. *Are Women Achieving Equity in Chemistry? Dissolving Disparity and Catalyzing Change*; Marzabadi, C. H., Kuck, V. J., Nolan, S. A., Buckner, J. P., Eds.; ACS Symposium Series 929; American Chemical Society: Washington, DC, 2006; pp 1–146 and references therein.
9. Raber, L. R. Academic hiring of women. *Chem. Eng. News* **2009**, *87* (23), 9.
10. Golde, C. M.; Dore, T. M. *At Cross Purposes: What the Experiences of Today's Doctoral Students Reveal about Doctoral Education*; Survey Initiated by the Pew Charitable Trusts; Philadelphia, PA, January 2001.
11. Goulden, M.; Frasch, K.; Mason, M. A. *Staying Competitive: Patching America's Leaky Pipeline in the Sciences*; The Center for American Progress: Washington, DC, November 2009.
12. Kuck, V. J.; Marzabadi, C. H.; Nolan, S. A.; Buckner, J. P. Analysis by gender of the doctoral and postdoctoral institutions of faculty members at top-fifty ranked chemistry departments. *J. Chem. Educ.* **2004**, *81*, 356–363.
13. Kuck, V. J.; Marzabadi, C. H.; Buckner, J. P.; Nolan, S. A. A Review and study on graduate training and academic hiring of chemists. *J. Chem Educ.* **2007**, *84*, 277–284.
14. Nolan, S. A.; Buckner, J. P.; Marzabadi, C. H.; Kuck, V. J. Training and mentoring of chemists: A study of gender disparity. *Sex Roles* **2008**, *58*, 235–250.
15. Watt, S. Assessing the effects of (post)doctoral-level experiences on graduate students' and postdoctoral associates' professional needs and career choices in a large chemistry department. Manuscript in preparation.

16. Page, S. *The Difference: How the Power of Diversity Creates Better Groups, Firms, Schools, and Societies*; Princeton University Press: Princeton, NJ, 2007.

17. McLeod, P. L.; Nobel, S. A.; Cox, T. H., Jr. Ethnic diversity and creativity in small groups. *Small Group Res.* **1996**, *27*, 248–264.

18. Watt, S. *Fostering the Co-Curricular Success of University of Michigan Chemistry Graduate Students and Postdoctoral Associates*; Internal Report to the University of Michigan Chemistry Faculty; University of Michigan: Ann Arbor, MI, 2010.

19. Grunert, M. L. Purdue University, East Lafayette, IN. Personal communication, 2010.

Chapter 3

Adapting Mentoring Programs to the Liberal Arts College Environment

Kristin M. Fox,[*,1] Catherine White Berheide,[2]
Kimberley A. Frederick,[3] and Brenda Johnson[4]

[1]Department of Chemistry, Union College, Schenectady, NY 12308
[2]Department of Sociology, Skidmore College, Saratoga Springs, NY 12866
[3]Department of Chemistry, Skidmore College, Saratoga Springs, NY 12866
[4]Department of Mathematics, Union College, Schenectady, NY 12308
[*]foxk@union.edu

Recent work indicates that mentoring of both tenure-track and tenured STEM women faculty is important for their success. Surveys at Union and Skidmore Colleges have shown that faculty rising through the ranks agree that they need more information about the tenure and promotion process and that having a mentoring program is important to them. Because the development of mentoring networks is considered more beneficial than having a single mentor, the overall goal of our project is to provide faculty with a variety of mentors who can share their successes and challenges. At small institutions such as ours, drawing from the population of both colleges for mentors is advantageous. As a complement to the pre-existing mentoring programs on both campuses, we have developed a series of mentoring opportunities mainly for STEM women faculty, including speed mentoring, receptions, discussion tables, peer mentoring, and a mentoring database. STEM women have found the opportunity to exchange information on achievements and challenges, both personal and professional, to be empowering.

© 2010 American Chemical Society

In the summer of 2008, Skidmore and Union Colleges received funding from the NSF ADVANCE program to adapt exemplary tools developed through ADVANCE Institutional Transformation programs at large research institutions to the climate and conditions at small liberal arts colleges. Skidmore and Union Colleges are both highly selective private liberal arts colleges of similar size in the Capital District of New York State that differ from one another in significant ways. Skidmore, co-educational since 1971, was originally a women's college that traditionally emphasized the arts and humanities. Over the course of the past decade, it has increased the role of the science, technology, engineering, and mathematics (STEM) disciplines in its curriculum. In contrast, Union is a formerly all-male college, also coeducational since 1970, that historically has had a strong natural science and engineering orientation—approximately 40% of its students major in the lab sciences and engineering. Thus, the two institutions bring different experiences and strengths to the project, and therefore tools developed for this project are expected to have broad applicability to a wide variety of liberal arts institutions. (More information about the SUN NSF-ADVANCE program can be found at http://sun.skidmore.union.edu/.) One important goal of our NSF project involves providing resources and support, including mentoring, that will help assistant and associate professors advance in rank.

By virtue of their smaller size, academic departments in liberal arts colleges often have only one faculty member in a particular subfield. Therefore, there is a much lower likelihood of women faculty finding a mentor with similar scholarly interests, let alone one of the same gender, than at a research university. Cross-institutional relationships become critical when there is only one woman in a department who teaches a particular topic or does research in a particular subfield. They are also, as Gibson notes (1), a way to "avoid some of the political constraints of being mentored in one's academic department."

To address the potential lack of a suitable mentor within the department, we have formed the SUN (Skidmore-Union) Women Faculty Network, a cross-institutional network among women faculty in STEM departments at the two colleges. Our goal is two-fold: first to facilitate finding a mentor(s) among the STEM women at the home institution who can provide campus-specific forms of mentoring, especially related to tenure and promotion processes, and second to facilitate finding a mentor(s) among the STEM women at the partner institution who can provide discipline- and preferably sub-discipline-specific forms of mentoring, especially related to teaching and research. This approach has enabled women faculty to go beyond their own departments and institutions to form relationships with others who might provide the pedagogical expertise, similarity of research interests, and/or the psychosocial support they uniquely require. This partnership model of building cross-institutional linkages to enhance mentoring and development opportunities for STEM women faculty may be appropriate for other small liberal arts colleges.

In this paper, we describe the rationale for the mentoring program we are designing. This design is informed both by the mentoring literature and by survey and focus group data from Skidmore and Union faculty. We also provide an overview of the specific components of our mentoring program and our colleagues' responses to them. We begin the paper with information about the mentoring climate at our institutions, describing mentoring programs already in place at the beginning of the grant period and the survey and focus group information about mentoring that we gathered early in the grant period.

Mentoring at Skidmore and Union: Institutional Programs and Faculty Attitudes

Pre-Existing Mentoring Programs

Both campuses have had various mentoring efforts over the years, but here we will focus only on the efforts that were in place at the two institutions just before the grant was received. At Union College, the Union Coalition for Inclusiveness and Diversity (UCID) has recently established a program for all pre-tenure faculty members, visiting faculty, and recently appointed lecturers. It provides them with the opportunity to have a mentor from outside their own department to complement the mentoring they receive from colleagues within their department. Participation is completely voluntary. Junior faculty are not assigned a particular mentor but, instead, may choose anyone from a diverse pool of available mentors. Junior faculty members may change mentors at will.

Another ongoing program at Union is part of new faculty orientation and consists of three sessions each year. The first, before classes begin in the fall, is an all-day event focusing on teaching resources and policies. The second and third are made up of a dinner, a short talk, and a one-hour panel presentation and discussion. The second focuses on faculty scholarship and a member of the Grants Office gives the short talk. The third focuses on service to the department, college, and profession with a short talk about faculty advising. New department Chairs also receive a half-day orientation.

At Skidmore, the new faculty orientation program, coordinated by the Assistant Dean of Faculty and Director of the First Year Experience, has been reformulated as a new faculty learning community consisting of all first year faculty and several tenured faculty mentors. Interested second-year faculty are also invited to participate. Before the beginning of the semester, there is a day-long orientation program that focuses on practical faculty issues. After the initial meeting, the faculty learning community meets monthly for informal discussions about any issues of interest which may include classroom management, interpretation of student evaluations, and balancing work/life issues. The faculty mentors also take the new faculty out to lunch or host dinners at their homes.

Focus Group and Survey Data Evaluating Pre-Existing Mentoring

During the first year of the project, we conducted focus groups and administered a climate survey to gather data on the current status of women faculty in the STEM disciplines at the two colleges. During the second year of the project, we conducted a mentoring survey. All three sources of data indicated that most faculty believed mentoring is important and provided suggestions of some ways in which mentoring at the colleges might be improved.

Focus Groups

Three focus groups were held at each institution, one for each rank (assistant, associate, and full professor), resulting in a total of 6 groups. A typical group contained 6-8 women representing the full range of STEM disciplines at each college at each rank. The focus groups provided a rich set of baseline data for a variety of purposes, especially the development of the climate survey.

The issue of a formal mentorship program arose in the focus groups with assistant professors. For example, one natural scientist commented:

> I would have liked to see more of a concrete mentorship program, because I think that it would have been helpful for me to have someone outside of my department to talk to. And that it would have been helpful for me to have a go to person within the department. I think I would have benefitted from that.

A natural scientist in a different department responded:

> I would agree with that. I find that I can go to the senior members of the faculty and ask them questions, but it's just me asking them questions. There's no sort of rapport or give and take in mentorship. And sometimes I feel like I'm just pestering them with questions, but I don't know how else to find out things that I need to know.

Similarly, the women associate professors in the STEM disciplines on both campuses felt disadvantaged by the lack of mentoring around promotion to full. Specifically, they reported not getting sufficient information about when a faculty member is ready to go up for promotion to full professor. For example, when discussing the issue of when to stand for promotion to full professor a woman associate professor in the natural sciences observed that, "There's not very good advice," her social science colleague responded, "I don't get any real mentoring about it. Whereas when I was junior, I got tons of mentoring." A female assistant professor in the social sciences even remarked on this problem when discussing mentoring in her focus group, commenting that, "There seems to be no mentorship at the next levels. I mean, full professor." Thus the focus group data reinforced our plan to target our mentoring activities at associate professors as well as untenured women to help STEM women at both levels progress successfully to the next rank.

Climate Survey

A climate survey was administered to all tenure-track and tenured faculty employed at the two colleges between March 15 and June 16, 2009. Out of the 341 tenured and tenure-track faculty to whom the survey was sent, 237 completed it, yielding a 70% response rate. Table I shows demographic information for the survey respondents. STEM women at both institutions had a 91% response rate leading to their overrepresentation in the sample.

Table I. Demographics of Faculty Respondents

Characteristic	Frequency	Percentage
College of Employment		
Skidmore College	122	52
Union College	115	48
Professorial Rank		
Professor	90	38
Associate Professor	95	40
Assistant Professor	50	21
Lecturer	2	1
Gender		
Female	118	50
Male	119	50
Discipline		
STEM[1]	139	59
Non-STEM[2]	98	41

[1] STEM fields include the Social Sciences (Anthropology, Economics, Political Science, Psychology, and Sociology) in addition to the Natural Sciences, Engineering, and Mathematics. [2] Non-STEM fields include the arts, humanities, and professional programs (Education, Management and Business, and Social Work).

The survey was designed in the fall of 2008 to measure dimensions of faculty work life by adapting items from existing climate surveys used at the University of Alabama, University of Illinois at Chicago, University of Michigan, University of Texas, Utah State University, Virginia Tech, and University of Wisconsin to a liberal arts college setting. Items that did not pertain to liberal arts colleges, such as teaching graduate-level courses, were deleted, and some items that were particularly pertinent for liberal arts colleges were modified or added. The survey consisted of 53 questions covering the following domains:

- Demographics
- Tenure and Promotion
- Equipment, Supplies, and Resources
- Department Climate
- Balance between Professional and Personal Life
- Overall Satisfaction with Work/Job/Campus
- Work Load
- Health and Well-Being

The discussion here focuses only on four items that were particularly pertinent to mentoring.

Table II. Climate Survey Data (in percentages)

Statement	1- Strongly Disagree	2- Disagree	3- Somewhat Disagree	4- Somewhat Agree	5- Agree	6- Strongly Agree	Total (N)
Colleagues give career advice/ guidance[1]	8.7	9.2	9.2	22.0	26.1	24.8	100.0 (218)
Senior Colleagues Helpful Toward Tenure[2]	8.7	8.2	10.5	20.5	23.7	28.3	100.0 (219)
Senior Colleagues Helpful with Promotion to Full Professor[3]	19.1	10.8	19.1	19.7	17.8	13.4	100.0 (157)

	1- Very Unimportant	2- Unimportant	3- Somewhat Unimportant	4- Somewhat Important	5- Important	6- Very Important	Total (N)
Faculty Mentoring Program[4]	9.1	8.2	7.7	24.5	25.0	25.5	100.0 (208)

[1] This question was prefaced by the general statement: "How much do you agree or disagree with the following statements about resources available to you?" and the statement was "I have colleagues or peers who give me career advice and guidance when I need it."   [2] The question was prefaced by: "Please indicate your level of agreement with the following statements regarding your personal experience going through the tenure process in your department." and the statement was "My senior colleagues are/were very helpful to me in working toward tenure."   [3] The question was prefaced by: "Please indicate your level of agreement with the following statements regarding your personal experience with the process of promotion to full professor in your department." and the statement was "My senior colleagues are/were very helpful to me in working toward promotion to full professor."   [4] The item was prefaced by: "Thinking about what you need to do your job as a faculty member more successfully, especially to help you attain tenure or promotion to full professor if you have not already, please rate the importance of the following programs or policies".

The climate survey included three questions specifically about colleagues' help with the respondent's career: a general one, one specifically about tenure, and another about promotion to full professor (which only associate and full professors were asked). In addition, the climate survey asked faculty to rank the importance of a faculty mentoring program.

According to Table II, while almost three-quarters (73%) at least somewhat agreed that they had colleagues who gave advice and guidance about their careers, over one-quarter (27%) at least somewhat disagreed. These frequencies did not differ by gender, but they did differ by discipline with faculty in STEM fields more likely to have reported having colleagues who provide guidance and advice. Table III presents means by discipline along with the results for an independent samples t-test of the difference of means.

Over one-quarter of the faculty (28%) strongly agreed with the statement that senior colleagues were helpful in working toward tenure, while one-quarter (24%) agreed, and slightly less than one-quarter (21%) somewhat agreed (see Table II). Over one-quarter of the faculty (28%), therefore, at least somewhat disagreed that their senior colleagues were helpful in working towards tenure. These frequencies did not vary by rank, tenure status, gender, or discipline.

Table III. Comparisons of Faculty on Two Measures of Mentoring (n=139 STEM and 98 Non-STEM; n=90 Professors and 95 Associate Professors)

Variable	M	SD	t	df	p
I have colleagues who give career advice/guidance			-3.21	216	.002
STEM	4.50	1.46			
Non-STEM	3.82	1.63			
I have colleagues who give career advice/guidance			2.56	216	.011
Tenured	4.07	1.56			
Non-tenured	4.71	1.50			
Senior colleagues helpful towards promotion to full			-2.97	155	.003
Associate Professor	3.07	1.58			
Full Professor	3.84	1.68			

Only 13% of the full and associate professors strongly agreed that senior colleagues were helpful in working toward promotion, 18% agreed, and 20% somewhat agreed. While these frequencies did not vary by gender or discipline, they did vary by rank with the associate professors more likely to somewhat disagree while the full professors were more likely to somewhat agree that senior colleagues were helpful in working toward promotion. Table III presents means by discipline along with the results for an independent samples *t*-test of the difference of means.

When the climate survey asked respondents to rate the importance of a faculty mentoring program, one-quarter (26%) rated it as very important, one-quarter (25%) as important, and one-quarter (24%) as somewhat important (see Table II). The remaining quarter (25%) of the faculty felt that a mentoring program was at least somewhat unimportant. Thus, there appears to be consensus that faculty would benefit from a mentoring program.

Mentoring Survey

A short follow-up survey examining issues related specifically to mentoring was administered electronically to all tenured and tenure-track faculty at Skidmore and Union Colleges in the spring of 2010. Faculty were asked questions in the following general categories:

- Where faculty found mentors (in their department, in another department at Skidmore/Union, at a previous institution, at another institution)
- Their degree of satisfaction with their current mentoring
- The barriers to mentoring
- Areas in which the mentor provides guidance: teaching, student interactions, scholarship, support and guidance, work/life balance, institutional politics.

The response rate to the survey was 144 out of 353 (41%). In total, 46% (65) of the respondents were men and 54% (75) women, of whom 22% (31) were tenure-track and 78% (112) were tenured. Finally, 20% (28) were assistant professors, 43% (61) were associate professors, and 37% (53) were full professors.

Over half (56%) of the respondents indicated that they were satisfied with the mentoring they were currently receiving and almost two-thirds (63%) reported that they had found a good mentor in their department at some point in their career. The percentages of faculty who were satisfied did not differ significantly between tenured and tenure-track faculty. About 25% of full professors, however, replied "not applicable" to the question about current mentoring and noted in the comments section that they feel that they no longer need mentoring and instead provide mentoring.

When asked various questions about why they did not have a mentor, there was very little difference between tenured and tenure-track faculty. Only 10% of tenure track and 20% of tenured faculty agreed that they lacked the opportunity to meet mentors, and 16% in both ranks indicated that they lacked the opportunity to

34

develop mentoring relationships. Interestingly, one-quarter of tenure track (26%) and tenured faculty (27%) indicated that they would prefer to get mentoring in ways other than the traditional one-on-one senior mentor to junior mentor format.

The data we have gathered via the surveys and focus groups described above is being used to inform our approach to mentoring as our programs continue to develop. We learned that we need to continue to improve mentoring of pre-tenure faculty, and that post-tenure faculty also are in need of networking opportunities to improve access to information about being successful in their careers. The next section will list the mentoring/networking strategies we have developed and our plans for modifying them based on our experiences and the information we have gathered from focus groups, surveys, and assessments administered after each mentoring/networking event.

The Skidmore Union Network Mentoring Program

Our Approach

According to de Janasz and Sullivan (2), the traditional model of faculty being "guided throughout their careers by one primary mentor, usually the dissertation advisor" is no longer appropriate. They argue that faculty are better served by developing "multiple mentoring relationships across their academic career." Kirchmeyer (3) concurs, concluding that it is important to have an "entire constellation of developers performing functions important for protégé advancement," particularly since she found that working with developers who are outside of the mentee's institution resulted in more scholarly publications. Cawyer, Simonds, and Davis (4) also recommend fostering "informal mentoring from multiple faculty members."

In addition to having multiple mentors, other research finds peer mentoring effective. For example, Files et al. (5) found that a facilitated peer mentorship program for female medical faculty led to an increase in publications as well as promotion in rank. As a result of her research, Wasburn (6) recommends a strategic collaboration approach to mentoring that combines peer mentoring with developmental networks by matching two full professors with a peer group of three to five assistant or associate professors. Our approach seeks to foster a "constellation of developers" by providing opportunities for both vertical (across ranks) and horizontal (between individuals of the same rank) mentoring.

Information collected in the focus groups, the SUN climate survey, the mentoring survey, and informal conversations with assistant and associate STEM professors support our implementation of this approach to mentoring. This broader approach to mentoring is based on the principle we articulated in our NSF proposal that career development benefits from multiple sources, from junior colleagues as well as senior colleagues, from faculty in the same department as well as those from other departments, from faculty at the same institution as well as those from other institutions.

In short, as is true for other occupations, good mentoring plays an important role in a faculty member's career by enhancing his or her professional growth. Mentoring serves an especially critical role in the advancement of women faculty, perhaps even more so for those in traditionally male-dominated fields such as the STEM disciplines. Yet, women report low levels of mentoring and other forms of developmental relationships (6–8), perhaps because there is a shortage of female mentors, particularly women in the highest ranks. In addition, the reality of higher teaching loads, higher expectations for service, and heavier student advising responsibilities at liberal arts colleges compared to research universities results in less time for faculty mentoring or professional development by faculty of both genders. In response to our workplace climate survey, women faculty at Skidmore and Union Colleges reported significantly greater time pressures (M=22.03) than their male colleagues (M=19.29), a difference of 2.74 on a scale ranging from 5 to 30 ($t_{(224)}$=3.71, p<.001). For multiple reasons, therefore, faculty at liberal arts colleges, particularly women faculty, may not receive mentoring or may only receive it from a smaller number of people. This lack of mentoring is especially unfortunate because it appears to be more important to women faculty. It has a powerful effect on their perceptions of positive relationships with colleagues which in turn is strongly associated with their job satisfaction (7, 9–11). According to Gibson (1), mentoring affirms their self-worth as teachers and scholars, leads them to feel they are not alone, provides a feeling of connection, gives female faculty a sense that someone cares about their success, and situates them in an academic environment that supports their success.

We have sponsored a wide variety of activities in which mentoring is a component. These range from more formal activities such as speed mentoring to less formal interactions such as receptions. Our goal is to provide faculty throughout the ranks with many opportunities to get to know a variety of STEM women at their own campus as well as the other campus so that they feel that they have a network of people who can be role models, mentors, people to bounce ideas off of, or whatever seems desirable. Below we describe each of these types of activities, providing some information about how to set them up as well as representative feedback from participants.

Our Activities

Speed Mentoring

The speed mentoring program was adapted from Georgia Tech's ADVANCE initiative using materials developed by the University of Kentucky (12). In speed mentoring, tenured STEM women faculty (mentors) provide individual guidance to pre-tenure STEM women faculty (protégés) during an hour-long session that is organized like a speed dating event. Individual meetings between mentors and protégés are short, and participants are matched up in advance by the organizers. Our first event, in April 2009, was for STEM women at both institutions and occurred at Skidmore College, where each protégé met for three minutes with five different mentors. At the conclusion of the event, we had a reception to allow for longer conversations. An anonymous survey conducted after the event showed that 84% of women agreed with the statement that "I was able to meet women working at Union and Skidmore College whom I would not have had an opportunity to meet otherwise" and 79% agreed with the statement that "I would attend another speed mentoring event." The major component we planned to change was the meeting time (3 minutes), since 74% of the participants thought that it was "too short."

The second event was at Union College in Fall 2009, and involved sixteen Union STEM women. It was identical to the first event, except that the meeting times were extended to four minutes. Responses were similar to those in our first meeting, but 60% of participants agreed that the meeting times were "just right," with 30% agreeing they were "too short" and 10% "too long." It seems, therefore, that 4 minute meetings work better than 3 minute meetings at our institutions. Below, we include some comments from speed mentoring participants:

I thought the event ran very well. It was well organized and the number of mentors present was impressive. In an indirect way, just the number of mentors interested indicated support/interest in the more junior faculty and that they felt this was something important to contribute to. I would hope that as I progress through my career I would be able to help junior faculty establish themselves in ways similar to those that the SUN program aims to achieve. It was a great event and I hope there will be another one.

Thought it was wonderful! I received alot [sic] of useful feedback from both potential mentors and colleagues. I am less anxious about my 'long-term' research goals at the moment. I also found being around women was much less intimidating than I even imagined.

It was a nice afternoon - long enough to make some new contacts but not so long as to detract from the workday. I'm very glad I went.

Overall, speed mentoring offers participants a chance to test out various mentoring relationships and/or to get advice from several different colleagues on a particular question. After adjusting the meeting time, participants were satisfied with the format.

Receptions/Meals with Speakers

In contrast to speed mentoring, which is a very formal type of interaction, our receptions are more informal events. Another component of the SUN grant involves inviting prominent women in the STEM disciplines to our campuses to give research talks. The goal of this part of the program is to provide both mentoring and faculty development opportunities. Women scientists from other institutions can share research findings, form contacts, and identify opportunities for research collaboration with women scientists at Skidmore and Union Colleges. Each speaker spends part of a day meeting with interested women from both campuses. In addition, the group events surrounding these lectures, such as meals and receptions, give women at Union and Skidmore a chance to meet with each other as well as with the visiting scientist.

Overall, informal feedback from these events has been very positive. Women scientists enjoy the opportunity to interact with each other on a professional level. In addition, several collaboration opportunities have developed with the speakers and between women at Union and Skidmore as a result of these events.

Discussion Tables

Feedback from the speed mentoring evaluation and informal conversations among women revealed a desire to have more time to talk about important issues and to bounce ideas off of each other. With this goal in mind, we developed topic-based discussion tables. At these events, women were divided into groups, based on interest, focused on either research, teaching, or service issues. After a casual lunch, discussion on a particular question was initiated. In Fall 2009 at Union College and Spring 2010 at Skidmore College, the topics were:

How do I carve time out of a busy schedule to focus on research?

What role should service play in my career?

How do I interpret student comments on my course evaluations?

The participants were divided into groups of 7-8 members. A facilitator who had specific experience in the area and a note taker were recruited prior to the event. The groups were mixed with women from all ranks participating. This event was opened up to women faculty across the entire campus and was not restricted to STEM women.

In an anonymous evaluation administered after the event, 95% of participants agreed that "The topics chosen provided for interesting discussion" and >90% of participants agreed that they would "attend another SUN lunch," "recommend this kind of event to my peers," and "encourage Union College to continue sponsoring this event." Participants commented:

It was very well organized and our time was efficiently used. My team leader was excellent as well in listening to all comments, then grouping into topics and leading the responses.

I liked this format a lot and thought the group size was perfect (not too big and small enough that everyone had a voice).

Suggestions were solicited for future topics for discussion. In addition to the desire of many to discuss one of the other topics in the future, they suggested the following topics: tailoring research for undergraduate students, work/life balance, and managing students in the classroom. Suggested changes were to make sure that we go around the room and introduce ourselves and to have more time so that people could discuss more than one topic. We plan to continue to run these events under the same format, but will make sure we do introductions in the future. To address the desire to have a chance to discuss the other topics, we plan to run two topic-based discussion table events at each institution during each academic year.

Peer Mentoring

The peer mentoring events were designed to be opportunities for STEM women at the same rank to discuss issues they face and to brainstorm strategies for success. Four events were held for tenure-track STEM women at Skidmore and two meetings were organized at Union, one for all tenure-track STEM women and the other for all STEM women at the associate level. At Skidmore the focus of the meetings has been on issues that are of common concern. At each meeting, the topic for the next meeting is selected. For example, one meeting was a celebration for faculty who had recently been awarded tenure. It provided an occasion for newly tenured STEM women to share their tips for success as well as answer questions from their junior colleagues. The topic for the next meeting will be about making wise service choices. At Union each group has met only once. In both cases there was good discussion and brainstorming of ideas to improve productivity. These events are still in the early stages. We plan to hold the discussions again this academic year before doing an evaluation of the effectiveness of this mentoring strategy.

Networking Database

To facilitate cross-institutional networking, we are in the process of compiling a database of available mentors and their expertise. The database will be available to STEM faculty at Skidmore and Union. The strength of this approach is that it allows faculty to self-initiate their mentoring relationships rather than being paired by a "match-maker". This approach also allows the mentor and mentee to self-define the extent of their mentoring relationship along a continuum from a single discussion to an on-going dialogue. We are encouraging faculty at all levels to consider being mentors as well as using the database to identify a mentor. The mentoring database will also facilitate the formation of mentoring alliances. This project is still in the early stages, but will be based on the very successful NSF-ADVANCE funded effort discussed by Karukstis, et al. elsewhere in this book. We plan to develop teams of four women. Each team will have two women from Union and two from Skidmore and will be composed of either social scientists or natural scientists, mathematicians, and engineers.

Conclusions

The climate survey indicated that faculty felt that a mentoring program is important and that faculty desired more guidance about the tenure and promotion processes, especially promotion to full professor. The mentoring survey revealed that only half the faculty were satisfied with the mentoring they were receiving, leaving considerable room for improvement. Respondents expressed interest in having more opportunities to meet mentors and to get mentoring in ways other than the traditional one-on-one relationships.

With all the various mentoring events we have held thus far, we have found that the women who attended enjoyed the experience of interacting with other female faculty. They found the opportunity to exchange information on achievements and challenges, both personal and professional, to be empowering. Often, it was difficult to get people to stop talking and bring the event to a close. Frequently people would offer suggestions for topics that could be discussed at the next such event. The challenge, however, has been to get people to attend the events. Setting aside time in an already busy schedule for mentoring could sometimes seem to be a poor use of time, especially for tenure-track faculty. We had women at both the associate and full professor levels at all of the events, distributed relative to their numbers in those ranks. Frequently, however, tenure track faculty were underrepresented relative to their numbers at our events. This distribution suggests that we have not made the case for mentoring being a beneficial activity to the tenure-track faculty, and that tenured faculty have more interest in mentoring than is often assumed.

We have been successful at establishing a variety of opportunities for women at all levels. While they have been very effective for post-tenure women, they have had more limited success for pre-tenure women. Pre-tenure women seemed to be most interested in events that provide a clearer link to what is necessary to achieve tenure. While many may not consider the new faculty orientation "mentoring", it does involve providing guidance from other faculty, and probably should be considered mentoring. Many senior faculty were also clearly not convinced that they continue to need mentoring. They were, however, enthusiastic participants in mentoring events, suggesting that tailoring mentoring activities to both groups is an essential feature of a strong mentoring program.

Plans for the Future

Our grant will be active for at least one more year. Currently we are planning to continue with the following next year: discussion tables, speed mentoring, and the informational sessions available for new faculty. In addition, we would like to add mentor/mentee training sessions, continue to develop our database, and use the database to set up alliances related to those discussed elsewhere in this book. We will also be developing plans to institutionalize the best mentoring strategies that we have developed. Please check our website for updates.

Acknowledgments

We would like to express our appreciation to Caroline D'Abate (Skidmore) who wrote the original mentoring section of the ADVANCE grant and generated many of the mentoring ideas, to Muriel Poston (Skidmore), Alice Dean (Skidmore), and Gretchel Hathaway (Union) who helped to plan and implement the mentoring activities, to Cay Anderson-Hanley (Union) who worked on the design and analysis of the climate survey, to Lisa Christenson (Skidmore), who worked on the mentoring survey, and to undergraduate research assistants, Tara Kelley and Joelle Sklaar (Skidmore), who worked on analyzing the results of the climate survey. We would also like to thank our other Co-PIs and senior personnel for their work on the SUN ADVANCE project, Barbara Danowski (Union), David Hayes (Union), Holley Hodgins (Skidmore), Monica Raveret Richter (Skidmore), Cherrice Traver (Union) and Suthathip Yaisawarng (Union). In addition we would like to thank Beau Breslin (Skidmore) for partnering with the SUN program to develop and deliver mentoring activities. This project is supported by the National Science Foundation under Grant Numbers 0820080 and 0820032, the Skidmore-Union Network (SUN) Committee, and the Skidmore College Summer Collaborative Research Program. Any opinions, findings, and conclusions or recommendations expressed in this material are those of the authors and do not necessarily reflect the view of the National Science Foundation, the SUN Network Committee, or Skidmore and Union Colleges.

References

1. Gibson, S. K. *J. Career Dev.* **2004**, *30*, 173–188.
2. de Janasz, S. C.; Sullivan, S. E. *J. Vocat. Behav.* **2004**, *64*, 263–283.
3. Kirchmeyer, C. *Hum. Relat.* **2005**, *58*, 637–660.
4. Cawyer, C. S.; Simonds, C.; Davis, S. *Int. J. Qual. Studies Educ.* **2002**, *15*, 225–242.
5. Files, J. A.; Blair, J. E.; Mayer, A. P.; Ko, M. G. *J. Women's Health* **2008**, *17*, 1009–1015.
6. Wasburn, M. H. *Mentoring Tutoring* **2007**, *15*, 57–72.
7. Settles, I. H.; Cortina, L. M.; Stewart, A. J.; Malley *J. Psychol. Women Q.* **2007**, *31*, 270–281.
8. Smith, J. W.; Smith, W. J.; Markham, S. E. *J. Career Dev.* **2000**, *26*, 251–262.
9. Ambrose, S.; Huston, T.; Norman, M. *Res. Higher Educ.* **2005**, *46*, 803–830.
10. Bilimoria, D.; Joy, S.; Liang, X. *Hum. Resour. Manage.* **2008**, *47*, 423–441.
11. Bilimoria, D.; Perry, S. R.; Liang, X.; Stoller, E. P.; Higgins, P.; Taylor, C. *J. Technol. Transfer* **2006**, *31*, 355–365.
12. Ready…Set…MENTOR! A Speed Mentoring Toolkit. The President's Commission on Women, University of Kentucky. http://www.uky.edu/PCW/Speed%20Mentoring%20Toolkitfinal.pdf.

Chapter 4

Initial Impacts of an NSF ADVANCE-IT Award to Rutgers University from the Viewpoint of the Camden Campus

G. A. Arbuckle-Keil[*,1] and D. Valentine[2]

[1]Department of Chemistry, Rutgers, The State University of New Jersey, Camden, NJ 08102
[2]Office for the Promotion of Women in Science, Engineering and Mathematics (SciWomen), Rutgers, The State University of New Jersey, Piscataway, NJ 08854
*Arbuckle@camden.rutgers.edu

Rutgers, The State University of New Jersey, received an NSF ADVANCE Institutional Transformation (IT) grant in fall 2008. The goals of the RU FAIR (Rutgers University for Faculty Advancement and Institutional Re-imagination) project are to 're-imagine' mechanisms to support the participation and advancement of women in science, social science, engineering, and mathematics (SEM) on the three campuses. Senior SEM women faculty (RU FAIR Professors) were selected from the three Rutgers campuses through a competitive application process to organize and implement grassroots solutions for women in the SEM fields. These faculty are essential to enhancing communication within such a geographically and structurally complex multi-campus institution. The goals of the project are: to improve recruitment and retention initiatives; enhance communication between schools and campuses; encourage interdisciplinary research; improve visibility of women faculty; and to augment resources for dual-career families and families with children. The RU FAIR Professors on each of the Rutgers campuses share similar goals, though the unique institutional qualities of each campus shape the implementation and impact of the initiatives. In this chapter, we present initial activities and approaches to institutional

© 2010 American Chemical Society

transformation from the perspective of the Camden campus, a predominantly undergraduate institution (PUI) within a complex research-intensive university system.

Introduction

Rutgers University is a multi-sited university system that functions as the State University of New Jersey. It is the sole university in the United States that is a colonial college, a land-grant institution, and a public university (*1*). Rutgers is also a member of a select group of research-intensive institutions elected to the American Association of Universities (AAU). Understanding the structure, organizational complexity, and rich history of Rutgers provides a context against which the climate for female and minority faculty's full participation can be assessed. This also allows us to document the impact of the NSF ADVANCE Institutional Transformation grant on the university as a whole and on each campus individually.

Three campuses—Camden, Newark, and New Brunswick—constitute Rutgers University. New Jersey is a small densely populated state (ranking 46th in land area); Rutgers' three campuses are geographically dispersed across the state. Newark is located in the northern half of the state in New Jersey's largest city, directly across the Hudson River from New York City, connected by rail to New York and New Brunswick. Rutgers-New Brunswick, the administrative hub and largest campus in the Rutgers system, is located in central New Jersey. Camden is a city in the greater Philadelphia region with a rich history as an industrial and manufacturing center, though today it is often known for urban dysfunction, systemic poverty, and a high incidence of violent crime. The Camden campus is in the heart of the city's downtown, connected by public transportation to Philadelphia. Figure 1 shows the geographic dispersal of the Rutgers system across the state.

Countless seminars and other educational events are routinely held at the Rutgers-New Brunswick campus. The challenge for faculty in Camden is being able to devote sufficient time to take advantage of these programs. The distance between Rutgers-New Brunswick and Rutgers-Camden is approximately 60 miles. Travel between campuses is possible, but a round-trip commute time of approximately three hours must be taken into account in addition to time spent attending meetings, workshops or seminars.

A single university president oversees the entire Rutgers system, with chancellors at Camden and Newark and an executive vice president at New Brunswick fulfilling roles similar to "provost" at other universities. New Brunswick is home to the original colonial college, an all-male school that was first chartered in 1766 as Queen's College. In 1864, Rutgers expanded when it was chosen (over Princeton University) to be the state's land-grant school. The doors to higher education for women, however, remained closed in New Brunswick until the establishment of the New Jersey College for Women (later named Douglass College after its founder and first dean, Mabel Smith Douglass) in 1918.

*Figure 1. Map of the State of New Jersey Showing the Locations of the Three
Rutgers Campuses*

Rutgers College officially became a university in 1924, and by order of the New Jersey State Legislature, a public university in 1945. Rutgers-New Brunswick expanded in Central Jersey with the acquisition of land across the Raritan River in Piscataway, and Rutgers University expanded geographically in New Jersey when it merged with the University of Newark in 1946 and the College of South Jersey and the South Jersey Law School in Camden in 1950.

Higher education in Camden dates back to the 1920s when the South Jersey Law School was founded in 1926 and the two-year College of South Jersey in 1927. Both institutions were folded into the State of New Jersey university system as Rutgers sought to expand geographically to serve students in traditional arts and sciences programs, and institutionally, to include professional schools in law, business, and nursing, among others. Significantly, the first graduating class of Rutgers College of South Jersey, as the Camden campus was known in 1952, included four women along with 38 men (2). The institutional mergers began a process of shifting the demographic make-up of Rutgers from being a university for men to one for women as well.

The state university system envisioned in the 1940s and 50s is today a multi-campus institution serving more than 54,600 students and employing 2803 full-time and 1347 part-time faculty (3). Sixty-eight percent of students and 74 percent of full-time faculty are based in New Brunswick. Camden, the focus of this paper, is a predominantly undergraduate institution (PUI) educating about 11 percent of Rutgers students and employing 9 percent of its faculty. Rutgers-Newark, which, like Camden, merged its law school and arts and sciences faculty with New Brunswick soon after the designation of Rutgers as the State University of New Jersey, is more than twice as large as Camden in terms of its student body. Figure 2 shows the population by campus for undergraduate and graduate/professional students across all three campuses.

The number of students enrolled in the professional schools of law and business at Camden factors significantly in the overall percentage of students enrolled in post-baccalaureate programs. The Graduate School (the first non-professional post-baccalaureate programs at Camden) was founded in 1981 offering Masters degrees in Biology and English. The Masters degree in Chemistry was first awarded in 2000. The ability to offer advanced graduate education, beyond terminal Masters degrees, at Rutgers-Camden was only recently approved. These PhD programs are inter-disciplinary and focus on unique areas of faculty expertise. For example, the School of Arts and Sciences programs that offer PhD degrees are Childhood Studies (first in the nation), Public Policy, and Computational and Integrative Biology. Rutgers-Newark offers PhD degrees in fourteen disciplines; its seven schools educate the highest proportion of part-time undergraduate, graduate, and professional students, though the number of part-time students at Camden is also significant as shown in Figure 2.

Chemistry Departments at Rutgers

Chemistry is taught on all three campuses at Rutgers as an Arts and Sciences discipline with specific chemistry degree options certified by the American Chemical Society (ACS) Committee on Professional Training. The ACS certified degree option is available on all three campuses. The Department of Chemistry and Chemical Biology in New Brunswick is part of the Division of Mathematics and Physical Sciences in the School of Arts and Sciences. It offers a chemistry major, a chemistry minor, and enriches highly qualified students with an undergraduate honors program. Graduate students can work toward an MS or a PhD degree. The department currently enrolls about 120 graduate students. In addition, there are approximately 35 postdoctoral researchers. Twenty-five percent of tenured and tenure-track faculty are women (n=9/36) as shown in Figure 3.

The Department of Chemistry in Newark also grants MS and PhD degrees and educates undergraduates who major or minor in the discipline. The department is composed of 14 full-time tenured or tenure-track faculty; of these, three are women (21.4%).

Rutgers-Camden offers an undergraduate major or minor and an honors program. The chemistry major has four options: traditional, certificate (ACS accredited), biochemistry, and chemistry-business. For graduate students, training is limited to the Masters degree at this time. In May 2010, ten undergraduate chemistry degrees were awarded and two students received the Masters degree. The majority of chemistry majors are female; this is consistent with the national trend of undergraduate women comprising more than 50% of the total undergraduate population. As of academic year 2009-10, four tenure-track faculty constitute the chemistry department; one is a woman.

Introductory chemistry courses are required not only for the students majoring in the discipline, but also for students concentrating in other science fields. Pre-medical students enroll in these courses as well. When the full enrollment of students taking chemistry at Camden is considered (approximately 300-400/semester), a faculty of four is clearly an indicator of an under-staffed department.

In the 1990s, the department consisted of six full-time tenured or tenure-track faculty, half of whom were women. A variety of circumstances led to the current status of one remaining female chemistry faculty member at Rutgers-Camden. One woman moved to the larger New Brunswick campus; one woman did not receive tenure; and one decided to relocate to another institution before the tenure evaluation. These are all normal circumstances in academics for both men and women. However, the state budget-related challenges in New Jersey have slowed plans to return to that former level of six chemistry faculty. A new assistant professor in physical chemistry (a male) was hired in the last academic year; he started in the fall of 2010. The number of tenure-track women in science in Camden continues to be low, particularly when compared to the national pipeline of women receiving doctorates in SEM disciplines (4).

Figure 3 shows the percentage and number of women in the three chemistry departments at Rutgers. These data are compared with the reported percentage of women doctorates employed as tenured or tenure-track faculty at two- and four-year educational institutions in the United States (dotted line close to 17%) (5) and with the percentage of women earning doctoral degrees in 1998 and 2007 (dashed lines at 32% and 38%) (4). The number below each bar is the number of female chemistry faculty in each department by campus. For the past several years in Camden, there has only been one woman, but the total number of faculty changed between 2008-09 and 2009-10, influencing the percentage.

The national pipeline of women doctorates shown in Figure 3 are for 1998 and 2007. The more recent doctoral graduates would be expected to be postdoctoral researchers in 2010, or new hires at junior ranks (tenure-track or non-tenured faculty). Women, who received their degrees twelve years ago, in 1998, would be expected to be the full-time tenured women today, with most at the associate level (accounting for years in postdoctoral and tenure-track positions). The third (dotted) line in the figure shows the number of women who were employed as full-time tenured or tenure-track faculty in 2007, the most recent year for which national data are reported. The difference between the percentage of doctorates

Percent Undergraduates

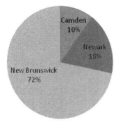

Percent Graduate/Professional Students

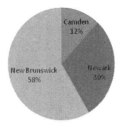

Percent Part-time Students

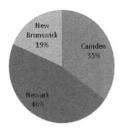

Percent Full-Time Faculty

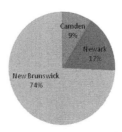

Figure 2. Four Pie Charts Comparing the Student and Faculty Populations at Rutgers University by Campus

earned and full-time academic employment in chemistry is due to a number of factors: employment in industry as professional chemists or leaving the field for other employment, for example (*6, 7*). The Rutgers chemistry departments are roughly on par with the national employment statistics and the Department of Chemistry and Chemical Biology in New Brunswick employs women as full-time tenured or tenure-track faculty at a level that is fairly close to the doctoral pipeline from 1998. The other campuses are close to national averages, but the Camden campus is not because of the low number of faculty. The bar for Camden, in fact, will drop to 20% in academic year 2010-11 due to the department growth to five full-time faculty with the addition of a male chemist.

Education and Support of Women at Rutgers

Historically, Rutgers University has a mixed reputation for educating women. For much of the twentieth century, educational opportunities for women remained fairly limited with the exception of Douglass College. By the 1950s, Douglass was the largest all-female public institution in the country and over the years, the eminence of the College in educating women and developing women's leadership programs has grown. Women scientists like Evelyn Witkin, a National Academy of Sciences geneticist who pioneered the study of DNA mutagenesis, taught women students at Douglass while their young male counterparts studied the same disciplines in courses and labs run by male faculty at Rutgers. When Rutgers College became co-educational in 1972, women had unprecedented

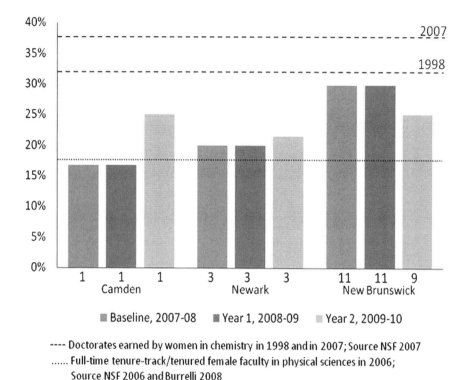

Figure 3. Graph Comparing Percent and Number of Female Chemistry Faculty
at Rutgers by Campus and Year

access to faculty across all of Rutgers. By 1976, 27 percent of all tenured and tenure-track faculty at Rutgers were women. They were concentrated in Douglass College where 43 percent of the 573 tenured and tenure-track women at Rutgers were employed in 1978 (8). Today, women constitute 36 percent of the tenured and tenure-track faculty at Rutgers University. Among the 34 public American Association of Universities (AAU) institutions, Rutgers ranked 5th with respect to the percent of full-time female faculty in 2007.

Both Douglass and Rutgers Colleges, along with Livingston and University Colleges, consolidated into a combined School of Arts and Sciences in 2007. Douglass now operates as a residential college offering programs to enhance women's learning and leadership development. The Institute for Women's Leadership (IWL), a consortium launched in 1991, shares that mission through its commitment to advancing women's leadership in education, research, politics, science, the arts, the workplace, and the world.

Recognizing the need to advance women's careers in the sciences, Rutgers University created the Office for the Promotion of Women in Science, Engineering, and Mathematics (Rutgers SciWomen) in 2006, appointing Associate Vice President and Professor II (9), Joan W. Bennett. Bennett is a noted fungal geneticist and a member of the National Academy of Sciences (along with genetist Evelyn Witkin, mentioned above, Bennett is one of only three Rutgers women to have been elected to the Academy). The SciWomen office supports women in science, social science, engineering, and mathematics at all levels at the university—from undergraduate and graduate women to faculty. The office also liaisons with people working in K-12 science and math education and with industry and other academic institutions in New Jersey. SciWomen organizes professional development workshops, sponsors lectures, outreaches to students and faculty, and celebrates women's accomplishments in science disciplines. The SciWomen website, for example, features two types of faculty profiles: professional biographies with links to personal and departmental websites and illustrated first-hand accounts of women's coming-of-age experiences as scientists or engineers. Dr. Kathryn Uhrich, a polymer chemist in the Department of Chemistry and Chemical Biology in New Brunswick and now Dean of Mathematical and Physical Sciences, describes her early interest in chemistry and the characteristics needed to be successful in academics (*10*). These profiles introduce people from around the world to the contributions of Rutgers' outstanding female scientists, foster a sense of community among Rutgers' women scientists, and enable young women to explore careers in science, social science, engineering, mathematics, and health/medicine.

SciWomen was successful in obtaining support from the National Science Foundation ADVANCE-IT program in 2008. Rutgers University for Faculty Advancement and Institutional Re-imagination (RU FAIR) (*11*) is a five-year award that aims to promote the participation and advancement of women in science, engineering, and mathematics (SEM) on all three campuses of Rutgers University. The goal of the RU FAIR ADVANCE program is to remove barriers to recruitment and retention of women faculty, to advocate for greater diversity in senior leadership positions, and to provide higher visibility to the achievements of Rutgers' women faculty in SEM disciplines. These goals can be summed as five core initiatives that drive institutional transformation and RU FAIR programs and activities: recruitment and retention; enhanced communication between schools and campuses; networking and liaisons that encourage interdisciplinary collaborations; improved visibility for women in science; and work/family issues. The mechanisms in place to achieve these goals include working with institutional partners to recruit and retain women faculty (RU InStride); institutional research on demographic indicators, RU FAIR mini-grant and life-cycle grant awards, and the RU FAIR Professorship program. Mini-grants awarded to faculty fund research or programs related to enhancing women's participation and advancement in science disciplines. Life-cycle grants support faculty at critical career junctures who are facing personal challenges. One key mechanism is the RU FAIR Professorship program which enables individual faculty to take on leadership roles in advancing women's participation in the sciences. They serve as university leaders who foster mentoring, promote diversity, facilitate communication among our geographically dispersed faculty, and mediate between faculty and administration.

The RU FAIR Professorship at Camden

The model for the RU FAIR professorship program originated from Georgia Tech's ADVANCE professorships (*12*), but the program at Rutgers has since evolved to fit the needs of faculty on three campuses and to address challenges and opportunities specific to Rutgers. Three professorships were awarded in March 2009 to senior faculty leaders based on a competitive application process and review by a selection committee that included the PI and co-PIs and the Internal Advisory Board of RU FAIR. First author, Georgia Arbuckle-Keil, is the RU FAIR professor at Camden, Judith Weis and Maggie Shiffrar share the RU FAIR professorship at Newark, and Helen Buettner is the RU FAIR professor (as well as grant co-PI) at New Brunswick.

The RU FAIR Professors encourage, organize, and implement grassroots solutions for women in the SEM fields specific to their campus needs. On the Camden campus, examples of activities include workshops, mentoring, and research on women's participation in SEM at Camden. The ACS symposium in March 2010 marked the one year anniversary of the RU FAIR Professorship and initial successes were presented. Additional events through fall 2010 are described herein. This conveys only the initial impacts of a five-year ADVANCE award; longer-term and enduring impacts of the grant will unfold in the years to come.

One event which took place during the first year was a grant-writing workshop to assist both male and female faculty with writing and receiving federal research support. The workshop, led by Catherine Duckett, one of the initial RU FAIR co-PIs, and who is now Associate Dean at the School of Sciences at Monmouth University, was well received and plans were developed to run the workshop as an annual event. To that end, two of Rutgers' experts on external grantfunding, Vice President for Research and Graduate and Professional Education, Michael J. Pazzani, and Camie Morrison, Director of Sponsored Research at Rutgers-Camden, led an RU FAIR-sponsored program in fall 2010 for faculty on navigating the process of successful grantsmanship. Pazzani, who served as NSF's Director of Information and Intelligent Systems (Division of Computer and Information Science and Engineering), brought years of experience on federal funding mechanisms as well as strategic insights to facilitate Rutgers' competitiveness for successful grantsmanship. Morrison delivered knowledge of grant funding operations from the Rutgers-Camden perspective. More than thirty faculty members filled the room for the half-day session. It was part of a larger professional development series that took place at Camden during fall 2010 (see below).

Another successful workshop sponsored by the Camden RU FAIR Professorship program was a two-day event designed to help women deal with multiple personal and professional demands. This event was facilitated by Julie Cohen, a professional coach and the author of *Your Work, Your Life, Your Way: 7 Keys to Work-Life Balance* (*13*). Fourteen faculty women from Camden attended these workshops and the feedback was overwhelmingly positive. The Work-Life balance workshop will be offered again in the upcoming academic year.

A series of professional development and leadership workshops modeled on the OASIS (Objective Analysis of Self and Institution Seminar) program, developed and implemented by Rutgers SciWomen for faculty and industry women at Rutgers-New Brunswick, was held in fall 2010 at Rutgers-Camden. The OASIS acronym provides a unique name for the leadership program, evoking the kind of safe place where women can be replenished for professional growth and leadership. Participants develop an understanding of themselves as leaders as well as gain insights into the context of gender and leadership within the institution. First author, Georgia Arbuckle-Keil, participated in one of the early versions of the OASIS program at Rutgers-New Brunswick prior to the awarding of the ADVANCE grant. This enabled her to network with women scientists from several departments and learn first-hand the benefit of discussing academic challenges with other faculty to acquire new viewpoints. Knowing senior faculty personally makes it easier to contact them for assistance. For example, while writing a proposal to NSF for instrumentation, she realized that the proposal would be strengthened by including a materials scientist from New Brunswick. Due to proposal deadline constraints, the information was required almost immediately. Georgia was delighted and grateful that this well-known female scientist would take the time on short notice to provide documentation that her research group would utilize this instrumentation in Camden. The on-going RU FAIR programs at Rutgers-New Brunswick provide opportunities for Camden and Newark faculty to network with New Brunswick scientists. Since travel is not always possible, nor desirable, the first series of OASIS-style workshops were held at Rutgers-Camden in fall 2010.

The OASIS program in Camden included sessions on leadership development, writing, grantsmanship, and faculty-to-faculty coaching (co-mentoring). Department chairs were asked to nominate women faculty to participate. The goal was to recruit twenty-five women who would commit to attending all four half-day workshops as well as to meeting informally between sessions for co-mentoring. Priority was given to women scientists, but due to the small number of women in the physical sciences in Camden, women from other disciplines (social science, public policy, library science, nursing, business and law) were invited as well. In addition, five women in natural science fields were recruited from other local institutions of higher education, thereby serving a broader pool of science women while showcasing Rutgers-Camden as a leading partner for women's professional development.

To enable a network for faculty-to-faculty coaching, the larger group was broken up into smaller co-mentoring groups, each consisting of approximately five members who met either on campus or via conference call. Attempts were made to diversify each co-mentoring group as much as possible; for example, each group had one non-Rutgers faculty member. If two faculty from the same department participated, they were placed in different co-mentoring groups. This ensured a rich and varied experience for the group participants.

Each workshop in the OASIS series included interactive components. For example, the leadership-style workshop involved time for "speed networking," a popular method for connecting one professional to another. The idea is that two faculty who do not know each other share basic information for five-minutes. At that point, new faculty pairs are formed for another five-minute interval. This continues for as long as new pairs and time permit. Five minutes is about the length of time to realize that one would like to get to know the other faculty member better, and the process of creating lasting professional relationships continues over the course of the workshop sessions.

The OASIS participants have found the workshops to be informative and the networking opportunities enjoyable and professionally helpful. Current participants have already recommended to fellow faculty that they attend future OASIS workshops. The model at Rutgers-New Brunswick has been extended to include women from industry. This industrial support has helped off-set the operating costs, such as honoraria for workshop speakers and materials. In the future, the Camden OASIS may reach out to industry women in the local area who are working in chemistry and related bio-technology fields.

In addition to organizing workshops for faculty professional development, the RU FAIR Professorship program encourages research on the institutional climate for increasing women and minority faculty's participation and advancement in the sciences. Another project that Camden RU FAIR Professor, Georgia Arbuckle-Keil, spearheaded was a survey of male and female Camden faculty who had left the university. After obtaining Internal Review Board (IRB) approval, a Rutgers-New Brunswick sociology graduate student and RU FAIR graduate assistant, Crystal Bedley conducted phone interviews with faculty who were no longer at Rutgers-Camden. Interviews with faculty who may have received external offers and decided to remain at Rutgers, termed "stayer" interviews, are planned but have not yet been completed. "Exit" interviews were obtained from a small group of male and female physical science tenure-track faculty. The results were instructive, providing useful information about the experiences and perspectives of faculty members at Rutgers. Areas of concern these former faculty expressed included availability of resources, professional development, and promotion/tenure criteria. Specifically, interviewees noted that there is very little technical staff support available in the laboratory science departments. Faculty teaching laboratory courses are fully responsible for reagent preparation, testing/calibration of instrumentation, and laboratory clean-up. Limited student assistance is available via work-study, but this varies from semester to semester. Shared research instrumentation that might be available in a larger department must be located off-site and may also incur user fees. This is fairly common for a predominantly undergraduate institution (PUI). Maintaining sufficient progress in one's primary research to obtain tenure at a research university while simultaneously managing the heavier teaching responsibilities typical of a PUI, lack of graduate students and instrumentation, and other challenges, requires individuals to be extremely resourceful.

Faculty mentioned the need for research collaborations in order to advance their productivity. The university as a whole is a vast resource, but developing individual productive collaborations requires commitment from both researchers for the length of the project. The teaching responsibilities of the Camden faculty often preclude traveling to New Brunswick or Newark to develop these interactions, especially as junior faculty. Promotion criteria are clearly stated when faculty are hired: research, teaching, and service. Most department chairs of SEM departments clearly state that the primary tenure requirements for physical scientists are the demonstration of federal research support (grants) and the publication of papers in highly respected journals. Rutgers University, of which Camden is a part, is a research-intensive AAU institution. The faculty at Camden are proud to be part of an AAU Institution that includes many notable scholars in their disciplines. All tenure/promotion decisions are made by the same university-wide body: the Promotion Review Committee (PRC). Camden faculty do not want different review criteria that could lead them to be perceived as lesser scholars. However, the exit interviews reinforce the already strong consensus that the amount of time devoted to teaching at Rutgers-Camden has a negative influence on the quantity of scholarly productivity.

Isolation Challenges

The Camden College of Arts and Sciences has had a formal mentoring program for tenure-track faculty in place for several years. It is seen as a model for the university as a whole. SciWomen and the opportunities afforded by RU FAIR ADVANCE activities also seek to provide support for tenured faculty as they continue to advance through the professorial ranks. As there are currently no tenure-track or tenured women (other than Arbuckle-Keil) in the departments of chemistry, biology, or physics, the more pressing need is for successful recruitment of women. Although recent academic searches in these departments have included interviews of women, and in many cases, discussion of appointment, the searches ended without the addition of female faculty.

A recent article in Chemical & Engineering News (C&EN) provides some insight on this point. The headline of the article is encouraging: "Women Now 17% of Chemistry Faculty (*14*)." However, the graph reproduced here with permission from C&EN as Figure 4 documents a major concern. Results indicate that more women decide to accept appointments in chemistry departments that have six or more women, while departments with fewer than three women show a decline. The National Academy of Science's report, "Beyond Bias and Barriers," reached a similar conclusion: When women constitute a "critical mass" in a specific department, they "join together to press for improvements in policies" (*15*). The recently published study by Geraldine Richmond: "Is the Academic Climate Chilly? The Views of Women Academic Chemists," notes that "women are underrepresented in academia in comparison to their representation in the chemistry field as a whole." (*16*) The significance of this study is that it represents information from more than 250 women chemists from universities across the country who attended workshops sponsored by the Committee for the Advancement of Women Chemists (COACh). "Currently in the US there is a significant disparity in the recruitment and retention of women in the field of chemistry relative to their male counterparts, particularly at advanced degree levels." (*17*) The COACh workshops, usually held at national meetings of the American Chemical Society (ACS) or American Institute of Chemical Engineers (AIChE), provide professional development for women faculty in the chemical sciences. The leadership development workshops of SciWomen, including

OASIS, function to meet similar needs for Rutgers women, including providing local networking opportunities.

The data presented in Figure 4 agree with the experience at Rutgers-Camden where the department included three tenure-track or tenured women chemistry faculty in the 1990s versus the current situation of only one tenured female. The biology department in Camden currently has similar concerns with nine male faculty and no women; recent attempts to hire women have not been successful. One of the aims of the RU FAIR ADVANCE program is to try to overcome these obstacles.

Strategies for Success

Increased Communication

The opportunities afforded by the RU FAIR program have encouraged women faculty to network. The Work-Life Balance workshop and OASIS workshops in Camden brought women from various disciplines together. Attendees were encouraged by the opportunity to discuss the challenges of academic life with other women. A monthly networking lunch for female faculty in the sciences (both social and physical science) at Rutgers-Camden has decreased the feelings of isolation experienced by faculty in departments with only one woman. As previously noted, there are no tenure-track faculty in biology or physics. Currently, both the mathematics and computer science departments each have one woman. The networking lunches are presently organized and publicized by the RU FAIR Professor. Attendance at events varies. Only a few women attended some of the early networking lunch events. However, approximately twenty-five faculty regularly attended the four OASIS workshops. Competing time demands are given by the women as an explanation for missing an event. But they recognize the importance, identify with their colleagues' experiences, and try to attend as much as possible.

There are definite benefits to face-to-face interactions. However, professional time commitments on campus prevent faculty from always being able to meet in person at the main campus in New Brunswick. The RU FAIR Professor is a member of the internal advisory board for the NSF ADVANCE grant. Over the past two years, one meeting was held in Camden and one in Newark, but most meetings of these executive team and administrative sessions are held on the New Brunswick campus. Video-conferencing provides an alternative for some meetings that saves significant round-trip travel time, but works best after relationships have been developed so that all participants know and work well together. Attention must be given to include the participant who is not physically present at the meeting in the discussion.

An online course and project management system, Sakai (*18*) is used to facilitate communication between female faculty both at Camden and between campuses. Announcements of RU FAIR ADVANCE events are emailed regularly to women who have joined the SciWomen listserv.

For institutions without ADVANCE programs, workshops such as those offered by COACh provide opportunities to network with other female academic

ISOLATION DIMINISHES
Departments with six or more women increase as those with fewer than three decline

Number of departments

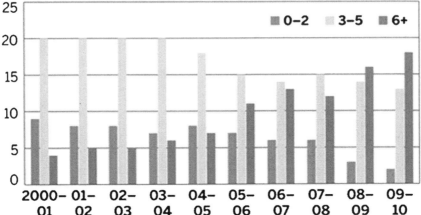

NOTE: Data are from the 33 universities that have appeared for the past 10 years on the National Science Foundation's list of the 50 biggest spenders on chemistry research.
SOURCE: C&EN surveys

Figure 4. Graph Comparing Number of Female Faculty in Chemistry Departments by Year. (Reproduced with permission from reference (14). Copyright 2010 ACS.)

scientists. Women are empowered to be leaders at their home institutions when they see that other faculty have similar experiences and are provided with tangible strategies to improve their leadership abilities.

Resourcefulness

Successful faculty at PUIs readily learn to be resourceful. Strategies range from the selection of the best undergraduate researchers, to meeting other colleagues who provide connections to scientists with instrumentation required for a specific measurement, to listening to the right department colleagues who have a realistic yet positive outlook and not allowing negative statements to hinder your research productivity.

Working primarily with undergraduate research students can be both challenging and gratifying. At a PUI, undergraduate student researchers are essential. The appropriate selection of talented students with enthusiasm for a specific research project can make the difference between acquiring publishable results and simply teaching students new laboratory skills. Although the teaching and training of undergraduates is its own reward, faculty must also produce publishable results if they expect to remain or advance in academia.

Early on, it is important to know which faculty members will provide assistance with issues that range from the mundane (which vendors provide the best discounts to the university), to the stomach churning (how to handle a disgruntled student). Helpful colleagues become your mentors and ease the anxieties of navigation through the waters as an untenured professor. Newly hired tenure-track faculty are encouraged by the Dean's office and their department chairs to meet a tenured faculty member from another department on campus. Networking with other faculty is generally useful, but for some discipline-specific information, colleagues from other neighboring institutions may be more appropriate.

Local sections of the American Chemical Society (ACS) can serve an important role. Some limitations of small department size, common to PUIs, can be overcome by networking with chemists at other regional institutions. Rutgers-Camden is part of the local Philadelphia Section of the ACS, which provides numerous networking and leadership opportunities. As an assistant professor at Rutgers-Camden, Arbuckle-Keil was asked to stand for election to the Board of Directors of the Philadelphia Section. The monthly meetings of the Board supplied ample opportunities to network with both local industrial and academic chemists. ACS members are a great resource. Faculty in small departments can network with chemists outside their institution, thereby obtaining both professional and personal support. The leadership opportunities at the local section led Arbuckle-Keil to chair the Philadelphia section in 2001; she currently serves as a Philadelphia section councilor and member of the Member Affairs Committee (MAC) at the national meetings.

Other professional societies can serve similar roles. The local section of the Association of Women in Science (AWIS) holds regular meetings in the Philadelphia area. If travel support is available, junior faculty should attend professional meetings to enhance visibility and improve their network.

Conclusions

The academic life of a women chemist at a predominantly undergraduate institution (PUI) within a research intensive institution is both challenging and rewarding. A wide variety of resources must be applied and specific needs change during the course of an academic career. The American Chemical Society is a rich resource. The ADVANCE award to Rutgers University provides new opportunities to encourage other women faculty to be successful at Rutgers University. The activities of the first year of RU FAIR ADVANCE have proved beneficial to women at Rutgers-Camden. As the RU FAIR project continues, we are hopeful that the number of women faculty in Camden will increase through improved recruitment and retention and that the leadership skills of senior faculty will be further developed.

Acknowledgments

The activities of the RU FAIR ADVANCE program in Camden would not have been possible without the support and assistance of many people. Acknowledgment is given to NSF for the support of this program (ADVANCE IT HRD 08-10978). Thanks as well to former and current PIs of this grant: Joan W. Bennett, Patricia Roos, Catherine Duckett, and Nancy Rosoff as well as Helen Buettner, Kathryn Uhrich, and Philip Yeagle; to the staff in the Office for the Promotion of Women in Science, Engineering and Mathematics (Natalie Batmanian, Christina Leshko, and Doreen Valentine); to Daitza Frydel, research assistant in Camden; Mary Gatta, internal evaluator for RU FAIR; and Crystal Bedley (the graduate student who conducted the exit interviews).

References

1. Rutgers History. Rutgers University Archives. http://www.rutgers.edu/about-rutgers/rutgers-history (accessed October 28, 2010).
2. Historic Overview of Rutgers-Camden. Rutgers University Archives. http://news.rutgers.edu/medrel/special-content/historic-overview-of-20100331 (accessed October 28, 2010).
3. Office of Institutional Research and Academic Planning FactBook 2009-10. http://oirap.rutgers.edu/instchar/factbook.html (accessed October 28, 2010).
4. Science and Engineering (S&E) doctoral degrees awarded to women, by field: 1998-2007, Table F-2. Women, Minorities, and Persons with Disabilities in Science and Engineering Report. National Science Foundation, Division of Science Resource Statistics, special tabulations of U.S. Department of Education, National Center for Education Statistics, Integrated Postsecondary Education Data System, Completions Survey, 1998-2007. http://www.nsf.gov/statistics/wmpd/pdf/tabf-2.pdf.
5. Women as a percentage of full-time full professors and full-time tenured/tenure-track faculty, by field of doctorate. National Science Foundation, Division of Science Resources Statistics, Survey of Doctoral Recipients: 1973-2006. In Burrelli, J. Thirty-Three Years of Women in S&E Faculty Positions; Info Brief Science Resources Statistics, NSF 08-308; July 2008. http://www.nsf.gov/statistics/infbrief/nsf08308/.
6. Sonnert, G.; Holton, G. *Who Succeeds in Science: The Gender Dimension*; Rutgers University Press: New Brunswick, 1995.
7. Hill, C.; Corbett, C.; St. Rose, A. *Why So Few?: Women in Science, Technology, Engineering, and Mathematics*; American Association of University Women (AAUW): Washington, DC, 2010. http://www.aauw.org/learn/research/whysofew.cfm.
8. Hawkesworth, M.; Hetfield, L.; Balliet, B.; Morgan, J. Feminist Interventions: Creating New Institutional Spaces for Women at Rutgers. In *Doing Diversity in Higher Education: Faculty Leaders Share Challenges and Strategies*; Brown-Glaude, W. R., Ed.; Rutgers University Press: New Brunswick, NJ, 2009; pp 137–165.

9. The designation of Professor II is reserved for those faculty in the University (usually already in the rank of professor) who have achieved scholarly eminence in their discipline. The standard for promotion to Professor II is significantly higher than that applied in promotion to professor.

10. Profiles of Women Faculty. http://sciencewomen.rutgers.edu/profiles (accessed October 28, 2010). This site was originally created by Catherine Duckett.

11. Rutgers University for Faculty Advancement and Institutional Re-imagination (RU FAIR). http://rufair.rutgers.edu (accessed October 28, 2010). Original PIs: Bennett, Duckett, Roos, and Rosoff.

12. Fox, M. F.; Colatrella, C.; McDowell, D.; Realff, M. L. Equity in Tenure and Promotion. In *Transforming Science and Engineering: Advancing Academic Women*; Stewart, A. J., Malley, J. E., Lavaque-Manty, D., Eds.; University of Michigan Press: Ann Arbor, MI, 2007; pp 170–186.

13. Cohen, J. *Your Work, Your Life, Your Way: 7 Keys to Work-Life Balance*; 2009. www.JulieCohenCoaching.com.

14. Raber, L. R. Women now 17% of chemistry faculty. *Chem. Eng. News* **2010**, *88* (9), 42–43.

15. *Beyond Bias and Barriers: Fulfilling the Potential of Women in Academic Science and Engineering*; The National Academies Press: Washington, DC, 2007.

16. Greene, J.; Lewis, P.; Richmond, G.; Stockard, J. Is the academic climate chilly? The views of women academic chemists. *J. Chem. Ed.* **2010**, *87* (4), 386–391.

17. Greene, J.; Lewis, P.; Richmond, G.; Stockard, J. COACh career development workshops for science and engineering faculty: Views of the career impact on women chemists and chemical engineers. *J. Chem. Ed.* **2010**, *87* (4), 381–385.

18. Sakai Course Management at Rutgers. http://sakai.rutgers.edu.

Chapter 5

Mentoring Pathways

A Small Wins Approach to Fostering Faculty Development

Toni Alexander,[1] Donna L. Sollie,[*,2] Victoria R. Brown,[3]
Daydrie Hague,[4] Overtoun Jenda,[5] Alice E. Smith,[6]
Daniel J. Svyantec,[7] and Marie W. Wooten[8]

[1]Department of Geology and Geography, Auburn University,
Auburn, AL 36849
[2]Women's Initiatives, Office of Diversity and Multicultural Affairs,
Auburn University, Auburn, AL 36849
[3]Department of Psychology, Auburn University, Auburn, AL 36849
[4]Department of Theatre, Auburn University, Auburn, AL 36849
[5]Office of Diversity and Multicultural Affairs, Auburn University,
Auburn, AL 36849
[6]Department of Industrial and Systems Engineering, Auburn, AL 36849
[7]Department of Psychology, Auburn University, Auburn, AL 36849
[8]College of Sciences and Mathematics, Auburn University,
Auburn, AL 36849
*sollidl@auburn.edu

Substantial scholarly work has documented the lack of
persistence among females in the academic STEM pipeline.
While intervention efforts over the last several decades have
increased the number of females earning graduate degrees, far
fewer of those remain in higher education throughout their
professional careers. In the past, this pattern was attributed
to overt discrimination, but today it is more commonly the
result of "tiny cuts" within female faculty careers resulting in
professional and social isolation. Through an examination of
the implementation of Auburn University's Strategic Diversity
Plan and the *ADVANCE Auburn* project, this chapter proposes
a "small wins" solution to improving success and retention
among female STEM faculty through a multifaceted approach
to faculty mentoring.

© 2010 American Chemical Society

Recognizing Tiny Cuts That Impede Women in Academic STEM

Over the past several decades, much scholarly effort has been devoted to assessing and addressing the continuing disparities that exist between the recruitment and retention of male and female faculty within higher education. While these disparities have in general decreased within the private professional sector, they remain firmly ingrained within higher education and are particularly pronounced within the STEM (Science, Technology, Engineering, and Mathematics) disciplines (*1, 2*). Efforts to address the dearth of female faculty members in STEM disciplines have traditionally interpreted the problem in terms of a pipeline from which there were few female scientists and engineers produced as a result of too few entering the disciplines. This perception, in turn, fostered support for intervention programs that would steer more women into the STEM disciplines and ultimately increase the number earning university degrees in those fields (*3*). Since the implementation of such efforts, the number of women earning Ph.D. degrees in the sciences has risen to half of all degrees awarded, but ultimately only 3 to 15 percent of tenured full professors in these disciplines are women (*4, 5*). Clearly, the academic pipeline is leaking and only a few of those women who enter it remain there throughout their professional careers (*3, 6*).

In contrast to the instances of overt bias and discrimination that were all too commonly faced by female STEM faculty in the past, Etzkowitz, Kemelgor and Uzzi attribute the loss of women within the academic STEM disciplines today primarily to "tiny cuts" inflicted upon their careers (*6*). For men, initial small advantages typically accumulate incrementally and can lead to significant influence and power with time. In contrast, the cumulative effects of small impediments may result in seemingly insurmountable barriers to professional academic success for women. Furthermore, the most influential and pervasive tiny cuts are those that interfere with the development of guiding professional networks that are an important source of socialization and mentoring.

The National Science Foundation (NSF) has recognized the difficulties in attracting and retaining women within academic STEM disciplines. As a result, NSF announced the establishment of a new funding program entitled ADVANCE: Increasing the Participation and Advancement of Women in Academic Science and Engineering Careers. The goal of ADVANCE is to accomplish institutional change by "transform[ing] academic environments in ways that enhance the participation and advancement of women in science and engineering." Since 2001, NSF has awarded over $135,000,000 to support ADVANCE projects at more than one hundred different institutions through two different types of grants (*7*). Grants for Institutional Transformation (IT) are awarded to institutions of higher education that undertake comprehensive projects aimed at transforming institutional policies or climate, with a subsection of IT-Catalyst grants directed at institutional self-assessment to uncover the need for transformation. Grants focused on Partnerships for Adaptation, Implementation and Dissemination (PAID) are designed to share information regarding gender issues as well as the results of institutional transformation projects, and are awarded to a broader range of institutions. ADVANCE-IT institutions consistently identify mentoring as a critical factor in the advancement and retention of women faculty in the STEM disciplines, and mentoring is a key element of the "small wins" approach that is the central driving force of the ADVANCE-PAID grant awarded to Auburn University in 2006.

A Small Wins Approach at Auburn University

The foundation for the transformation efforts at Auburn University was the Strategic Diversity Plan (SDP) of Auburn University (8). *ADVANCE Auburn* embraces the SDP as a guiding set of principles and members of the grant team have worked closely with the newly appointed Associate Provost of Diversity and Multicultural Affairs, who also serves as a co-principal investigator on the grant, to implement its goals and visions. The grant was initially written during a period of growing momentum in the STEM colleges, reflected in the hiring of seven women faculty in 2005, a record number in the College of Sciences and Mathematics and in the College of Engineering. Prior to the award of the grant, women accounted for only 12% of the faculty within the College of Sciences and Mathematics and 7% in the College of Engineering. As of Spring 2010, the percentage had increased to 17% in the College of Sciences and Mathematics and nearly 9% in the College of Engineering. These percentages of women faculty, however, are far less than the percentage of women faculty overall at Auburn University, which currently stands at 34%. In addition, these percentages are significantly lower than than the relative representation of female students at the University in STEM disciplines (55.2% of undergraduates and 38.2% of graduate students in the College of Sciences and Mathematics and 15.8% of undergraduates and 21.5% of graduate students in the College of Engineering are female). Obviously, there is a need to increase the number of women faculty in these underrepresented areas; but equally important are focused efforts to retain and promote these women faculty, with mentoring being a key factor in faculty development.

To further explore the needs of these new faculty and others like them, STEM networking sessions were held for women faculty in 2006-2007 where issues disproportionately affecting women faculty and their families, as well as means to deal with them, were discussed. Mentoring of junior STEM faculty and junior women faculty was intensified university-wide with an emphasis on providing the support and guidance needed for success, retention, and advancement.

As a land-grant institution, Auburn University is characterized by faculty who are deeply dedicated to educating students, conducting research, and serving the needs of the people of Alabama through extensive outreach. Such attitudes and achievements are attested to by Auburn being consistently ranked among the top 100 public universities (9). Auburn is a research institution, steeped in tradition, with strong alumni support. It was founded in 1856, was named a land-grant institution in 1872, and became co-educational in 1892. The institution was officially integrated in the 1960's; however, it continues to have difficulties in recruiting a diverse student body and faculty. As explained by Schein (10), the culture at Auburn has become so embedded in the people, processes and relationships that change is resisted even when demanded by a changing environment, including a changing gender and racial face in the workforce and student body. In 2001, the leadership of Auburn University officially recognized that increasing diversity in its faculty and student body would strengthen scholarship, provide a richer education for its students, make the institution more resilient in the new century, and more effectively serve the people of the state, region and nation. To that end, the University began addressing issues of diversity with renewed vigor through the creation of the Diversity Leadership Council, which developed the Strategic Diversity Plan (8). The basic tenets of the SDP are to foster a respectful and inclusive campus environment and to increase recruitment and retention of a diverse faculty, student body and supporting staff. The plan, accepted by the interim president in March 2005, includes a call for diversity efforts from all faculty, staff and students and charges the senior leaders of the University with the responsibility for guiding and monitoring "meaningful progress" (8).

Two additional signs of the university's desire to move diversity efforts forward were evidenced in the findings of two undertakings. A focus group of women and men faculty leaders overwhelmingly identified three main existing barriers to an inclusive faculty environment: 1) lack of clear policies to support balancing work and home life; 2) lack of official mentoring or support as women go through the tenure and promotion process; and 3) a campus culture that makes women feel unwelcome (11). Another indicator of collective interest in fundamental change occurred in the spring of 2005 when 100 participants in three follow-up strategic planning sessions convened by members of the Strategic Diversity Committee (12) repeatedly declared a need to attract more women and minority faculty and to improve work-life policies. Given this increased awareness, Auburn was poised for change when the NSF PAID grant was awarded. The grant was awarded the same month that one of the co-principal investigators at that time, and currently the principal investigator, was appointed to the newly created position of Assistant Provost for Women's Initiatives, a position that emerged as part of the Strategic Diversity Plan. In that role, she oversees faculty advancement initiatives, as well as the newly formed Women's Resource Center, the WISE (Women In Science and Engineering) Institute, and the *ADVANCE Auburn* Center, which was established as one of the grant objectives.

As an ADVANCE-PAID project, the programmatic goal of *ADVANCE Auburn* was the establishment of a "small wins" approach to influence lasting change in the culture and climate of the STEM disciplines at Auburn University. A small wins approach suggests that the overall transformation of an institution or workplace comes through incremental change – essentially, it recognizes that small changes can have widespread and long-term impacts (13). Rather than large-scale edicts from upper administration or radical organizational revolution, small wins practices that are implemented at the departmental, center, or college level result in greater buy-in from all administrative levels and ultimately more substantial institution-wide transformation (14–17).

The small wins approach is appealing because it allows for small or incremental costs (time and/or money) to return a substantial benefit to the institution, namely, an improvement in the climate for all faculty and greater retention of female faculty in the STEM disciplines. Auburn's grant had five objectives: 1) to establish the *ADVANCE Auburn* Center; 2) to assess the status of STEM women faculty and the climate within the STEM disciplines at Auburn University; 3) to develop a small wins cost/benefit model; 4) to select and implement small wins that have the highest benefit to cost ratio for transforming STEM disciplines and are most applicable to Auburn University; and 5) to disseminate the small wins cost/benefit model and implementation results. Two objective-related efforts have had a significant impact upon faculty mentoring initiatives: the administration of an AU faculty climate survey and a cost-benefit analysis of best practices employed at other ADVANCE-funded institutions. The findings from both of these endeavors have been used to develop and implement programmatic changes at Auburn.

Climate Survey Results on Mentoring

The faculty climate survey was designed and administered by the *ADVANCE Auburn* Center, in conjunction with the Office of Diversity and Multicultural Affairs, to help identify climate issues and impediments to the retention of female STEM faculty at Auburn University, as well as effective strategies to combat such barriers. While the resulting quantitative measures of evaluation suggested that mentoring of junior faculty was needed, participant responses to qualitative open-ended questions concerning departmental satisfaction offered additional insight. One female faculty member remarked, "Although my chair has been very helpful, there is no social support within my department, which makes being a new faculty member difficult." Another explained that "[m]ost of my disappointments with AU have to do with communication breakdowns between myself and my chair... the department was woefully unprepared for the influx of new faculty (there were 4 new hires this past year). We've all been pretty much left on our own to figure stuff out..." As these comments suggest, a lack of empathetic colleagues, inadequate coaching during professional transitions, and a lack of appropriate role models can serve to estrange female faculty from the rest of their department. Such subtle exclusion can occur socially and intellectually at both the departmental and college level as the following comment by another female faculty member illustrates:

> There is no connectivity between the program I am a part of and the department as a whole. There is little opportunity for intellectual stimulation outside my unit and no opportunities to network with other faculty members or programs within the department which would allow resource sharing, brain-trust capitalization, etc. Virtually no contact is made from the dept. head or dean from [our] college with our program.... We are very isolated and operate as an independent unit.

Ultimately, these combined tiny cuts foster an "emotionally draining" sense of both professional and personal isolation (*18*). This sense of isolation has negative consequences for both the individual faculty member and the institution as female

STEM faculty leave the university either due to inadequate professional academic success or by their own volition by seeking a job outside of academe.

The comments from the respondents in our climate survey reflect the typical struggles of female faculty to develop professional networks and become integrated within institutions at the departmental level, struggles that have been identified as all too common for women STEM faculty at other ADVANCE institutions. Tiny cuts ranging from a lack of social support and open lines of communication serve to alienate female faculty from their department and institutional colleagues and, in turn, deny them access to shared resources, intellectual communities, and the power structures of the institution. With a proper understanding of the local academic culture and guidance from those who have already succeeded along the professional academic path, however, female faculty can be encouraged, included, and retained. A female climate survey respondent noted the value of such assistance by explaining that "[M]y Department Head is a very organized leader who believes in shared governance, so I always feel I have a voice with him. I went though the Tenure and Promotion process 18 months ago and felt very prepared because of his foresight and guidance." As the comment suggests, mentoring can help set a new faculty member on the right path and play a key role in the success of female faculty members in departments where they are a minority.

Cost Benefit Analysis of Other ADVANCE Institutions

To complement the findings of the climate survey and further develop a small wins model, *ADVANCE Auburn* sought to understand how other ADVANCE-supported institutions have effectively implemented programs that might represent the small wins approach. A content analysis of ADVANCE program websites and published materials illustrated the primary approaches that have been employed by other universities. The most common initiatives were then grouped into general categories for evaluation: 1) Mentoring; 2) Family-friendly policies; 3) Training programs aimed at raising awareness of gender bias for various campus constituencies (students, faculty, search committees, etc.); 4) Department-wide workshops that highlight the scholarship of female faculty and provide guidance on improving departmental climate; 5) Departmental policies and resources that aim to improve the recruitment and retention of female faculty; and 6) Funding opportunities aimed at recruitment and retention of female faculty. A cost-benefit analysis was conducted to identify those practices that required the fewest resources and contributed the most to the improvement of the university climate and community. The directors of other ADVANCE-funded projects were asked to evaluate both the perceived costs and benefits of those practices that had been implemented at their institutions using a web-based survey instrument. Of the 72 ADVANCE grant principal investigators contacted via e-mail, 49 responded for a 68 percent response rate.

A cost-benefit ratio was calculated for each initiative by dividing the mean score for benefit by the mean score for cost. This ratio provided a measure for identifying programs that were the most impactful with the least cost, and would therefore be considered a small win. These programs have substantially aided in the career development of women in the STEM disciplines, but it should be noted that these interventions are of value to all untenured faculty. Moreover, the benefits are derived from changes in the organizational culture that have evolved from the motivated efforts of administrators and tenured faculty across the university. Of the 29 initiatives evaluated, mentoring programs represented over half of the most highly ranked practices employed at other universities. While the format of the mentoring program may vary from institution to institution, mentoring for different areas such as understanding the culture of the university or department, balancing work and family life, and providing insight into the promotion and tenure process were highly endorsed in our survey instrument by other ADVANCE grantees as being cost-effective and impactful. Creating programs that incorporate mentoring as a small win will not only aid female faculty development, but also improve the working environment for minority faculty, male faculty, and those from across the disciplinary spectrum alike.

Mentoring as a Pathway to Faculty Development

A recent article by de Janasz and Sullivan notes the limited amount of scholarly research related to mentoring faculty members in academia (*19*). The authors attribute this dearth to three main causes. The first is that new faculty members are presumed to have been fully prepared by their graduate studies. The implication then is that faculty members are assumed to have been mentored during their graduate studies, and to have maintained contact with that mentor. Neither of these assumptions is unreasonable, as graduate students typically choose or are assigned to a major professor under whom they are expected to master their chosen area of study. When hiring, academic institutions screen candidates carefully to ensure that the applicant has, in fact, mastered the area of study and is competent to teach. However, hiring institutions have no guarantee about the quality or continuation of that mentoring relationship. It may be erroneous to conclude that just because the major professor *can* mentor a graduate student that mentoring did occur, that the mentoring experience was beneficial for a future faculty member, or that the relationship will be maintained in this new environment. Additionally, many institutions expect junior faculty to establish an independent research program in order to demonstrate their capabilities and ability to function independently of their advisor. To this end, junior faculty may feel pressure to cut ties with their graduate advisor. These predicaments illustrate why a former graduate advisor cannot be solely responsible for mentoring new faculty members.

The second reason described by de Janasz and Sullivan for the lack of literature on mentoring in higher education is that the promotional ladder makes it difficult to identify appropriate mentors. Typically, new faculty begin as untenured assistant professors, advance with the award of tenure and promotion to the level of associate professor, and finally earn senior status as full professors. Despite what seems to be a fairly straightforward progression, new faculty do however vary considerably in their preparation and experience in teaching and research, with some arriving straight from graduate school, and others having had post-doctoral or professional experience. Additionally, each program, department, and academic institution has subtle nuances in environment and culture that impact the success of new faculty members. Tenured faculty colleagues can play a critical role in helping newcomers understand and navigate the unique departmental culture.

The third reason de Janasz and Sullivan give for the deficiency of literature on mentoring in academia is that some faculty, both junior and senior, perceive little need for mentoring (*19*). Senior scholars may feel that junior faculty members should be able to navigate the system on their own, referred to as the "sink or swim" model. Additionally, moving up the tenure ladder may only alter the duties of the faculty member in subtle ways. Unlike traditional organizations where a promotion often means a change in responsibilities, a promotion in the academic world is primarily a status change. An assistant professor is expected to teach, conduct research, and engage in service activities just as a tenured professor does. The seemingly static nature of expectations in academia may lead some to believe that mentors are not necessary (*20*). More recent evidence, however, indicates that support from senior faculty, department chairs, program heads, deans and other higher status academic professionals is crucial to the success of new faculty members (*21*). These findings suggest that mentoring is indeed needed in academia, and that if it were made available, it would be beneficial to those who wish to engage in a mentoring relationship. Furthermore, if mentoring is embedded into the academic culture, and if providing mentoring, guidance, and professional socialization is viewed as part of the responsibility of departments, colleges, and universities, junior faculty members might not fear the negative repercussions that could arise from acknowledging the need for such assistance.

Despite the lack of literature specific to higher education, a great deal of research has examined the impact of mentoring programs in other areas at both the individual employee level and the overall success of the organization. Kram defined mentoring as a developmental relationship between supervisors and subordinates, or among peers (*22*). However, this definition may not be the most applicable to mentoring in academic settings (*19*). It may be difficult to identify a single person who can serve as a mentor for all areas of interest, which is why a mentoring network consisting of multiple mentors for different areas is often advocated (*19, 22–24*), as is peer mentoring (*25*).

Many times department chairs or heads of a program area are expected to serve in a mentoring capacity; however, these supervisors may not be the most appropriate mentors due to personality conflicts, differing research specialties, or the added responsibilities of their roles that prevent them from committing significant time to a single faculty member. Moreover, the relationship between a mentor and protégé goes beyond that of a supervisor and subordinate. Ragins and Cotton found different mentoring styles for supervisors and non-supervisors: a supervisory mentor was able to provide more career-focused support, but not more social support, than a non-supervisory mentor (*26*). A mentor who is also a supervisor may have more direct access to career-advancement information that would be useful to the protégé than does a mentor who is not a supervisor. The reduced social support may be due to a hesitancy to engage in behaviors that may be seen as favoritism by other employees. Finally, it may be problematic to have a mentor who is in a position to formally evaluate the mentee. It is important to remember that a mentor and a supervisor may have very different roles, and that a chair cannot be assumed to serve as a mentor for an entire program or department.

There is a general consensus in the mentoring literature that naturally developing mentoring partnerships last longer and are deemed more successful than institutionalized mentoring partnerships (26). However, a recent study indicates that highly facilitated formalized mentoring programs can result in many positive outcomes, including more positive job attitudes through higher levels of job satisfaction and organizational commitment (27). This finding implies that it may be beneficial for universities and other academic institutions to implement highly structured mentoring programs to provide formal mentoring to junior faculty. A formal mentoring relationship may be abandoned if it is not beneficial, but it at least exposes junior faculty to the notion and importance of mentoring. This insight can encourage junior faculty members to seek their own informal mentoring relationships, which are likely to be longer lasting and more successful.

Mentoring can be very beneficial if the proper effort is put forth by both parties. Mentoring has been related to more clarity in a protégé's understanding of work responsibilities, as well as less conflict between the different areas for which a protégé may be responsible (28). Additionally, mentoring reduces perceptions of work-family conflict (29). Effective mentoring relationships have also been found to positively influence such tangible career outcomes as compensation, promotion, and reduced employee turnover, as well as improve overall job and career satisfaction (30, 31). In short, successful mentoring relationships produce more successful employees. The inference can then be made that successful mentoring relationships will result in more successful faculty, and potentially a better reputation for academic programs, departments, and institutions.

Faculty Mentoring as a Small Win at Auburn University

Given the lack of literature on mentoring specific to higher education, researchers must turn to other sources to gain insight on how mentoring relationships can be used. The need for mentoring as a small win to improve the university climate is evident from the response to Auburn's climate survey indicating professional isolation, as well as feedback from other institutions which suggest that mentoring is a cost-effective strategy for faculty development. There are several ongoing complementary mentoring programs and initiatives at Auburn University that support the goals of the ADVANCE grant, including programs provided by the Women's Initiatives Office, the WISE Institute, the Office of Diversity and Multicultural Affairs, and the Biggio Center for the Enhancement of Teaching & Learning, as well as programs within departments and colleges.

Mentoring programs at Auburn University in the Women's Initiatives Program include monthly informal networking opportunities for new women faculty to meet each other, as well as continuing male and female faculty. The program aims to provide a supportive network, possibilities for building research collaborations, and opportunities to learn from other faculty members about such topics as classroom issues, balancing work and family, and addressing departmental climate issues. Importantly, each of these events offers opportunities to interact with other faculty and learn about campus resources that can ease the transition into the faculty ranks, and develop both academic and personal social networks.

One very impactful brown bag lunch gathering included in the Women's Initiatives mentoring program offered a panel of three tenured faculty women representing several academic disciplines, who addressed the topic of "Questions I Wish I Had Had the Courage to Ask." The panelists openly shared their experiences during the pre-tenure years and identified topics and concerns that they felt negatively influenced their advancement. Concrete suggestions and practical advice were also provided, such as updating one's vita every 6 months, asking colleagues both within the department or college and within the field to provide feedback on manuscripts, and providing guidance concerning appropriate and effective ways to request assistance and resources from department chairs.

The Women's Initiatives Office has also established collaborations with a number of other campus departments and programs, including co-sponsoring programs that address concerns of women faculty with the Women's Studies Program. Among the most successful examples of this alliance was a brown bag luncheon that addressed the treatment of women faculty in the classroom, including such issues as student disrespect; expectations that women faculty should be more nurturing than their male colleagues; and the impacts of gender on teaching evaluations. The discussion surrounding these issues not only focused on the problems but also included tactics that more experienced women faculty had used to address such issues. Tactics suggested included: having a class discussion on appropriate classroom behavior towards the professor and other students; frowning at a disruptive student and briefly using silence as a reponse before continuing with the lecture or with classroom discussion; using student comments as a springboard for discussing the underlying issue represented in a negative or aggressive student comment; responding immediately to disrespect in the classroom by telling the student that you want to see him/her after class to discuss the behavior; or even using humor by saying "Did you really just say that?" If a faculty member feels that her gender is negatively impacting student evaluations, drawing the department head's attention to this concern is warranted, as is requesting assistance in addressing classroom issues and in increasing awareness within the department about bias in student evaluations.

Other campus organizations, including the WISE Institute, develop programming specifically aimed at the development and retention of female STEM faculty. WISE is governed by a Steering Committee that consists of women faculty and staff representing the Colleges of Agriculture, Sciences and Mathematics, Veterinary Medicine, Engineering, Education, Human Sciences, Architecture, Design & Construction as well as the Schools of Pharmacy, Nursing, and Forestry & Wildlife Sciences. The Steering Committee members serve as liaisons with their respective units and provide feedback as to the effectiveness and relevance of programming.

The Office of Women's Initiatives, the *ADVANCE Auburn* Center, and the WISE Institute co-sponsor a Speakers Series that features well-known women faculty from other universities who visit campus for two days, during which they present a research seminar, as well as additional talks on issues facing women in under-represented disciplines. These invited speakers also meet informally with graduate students and women faculty to address issues facing the advancement of women faculty members. Typically, there are two speakers each year, in the fall and spring semesters, and efforts are made to ensure that the speakers represent the departments and colleges that comprise the breadth of STEM disciplines. During this past academic year, Auburn co-hosted three speakers with the Colleges of Engineering and Veterinary Medicine. Overall, the assessment feedback from faculty and graduate students who attend these regular WISE events is overwhelmingly favorable and includes positive comments concerning such opportunities to meet other women in STEM and establish social support networks, as well as to learn how to develop professional networks and get involved in professional societies and conferences. Over the past two years, in addition to faculty from the Colleges of Engineering, Sciences and Mathematics, and Veterinary Medicine, the sessions have attracted increased numbers of faculty from other colleges (e.g., Agriculture, Forestry and Wildlife Sciences, Human Sciences, Nursing, and Pharmacy). Also, graduate students and postdocs have become more actively involved in the programs of the WISE Institute.

At times, efforts to develop professional networks have included support from beyond the Auburn campus. With support from the National Science Foundation, the WISE Institute recently collaborated with the *ADVANCE Auburn* Center and the Auburn University Graduate School to sponsor workshops for women faculty, graduate students and postdoctoral fellows in STEM disciplines. Entitled "COAChing Strong Women in Negotiation, Communication and Leadership" the workshops were organized and led by the Committee On the Advancement of women in Chemistry (COACh). Through the use of self-assessment, experiential learning, and role playing, the two workshops offered attendees the opportunity to develop communication skills crucial to women seeking professional academic success. Each of the two workshops focused upon the needs of differing constituencies, with one addressing the needs of faculty and the other for graduate students and post-docs. Having a separate workshop for graduate students and post-docs also highlights the importance of socializing women students and providing them with opportunities for professional development, as well as addressing the types of issues they may face as women in fields where they will be in the minority. In the workshop evaluations, both groups reported feeling empowered, more capable of negotiation, and less isolated. One faculty member planned to use the skills and tactics from the workshop to "apply to the negotiation process in terms of lessening committee work and taking more credit for grants obtained or in the process of writing," and another planned to "negotiate with my dean for additional resources for my office." Sessions like these remind females at all levels that they are not alone in the issues that they face, and offer exposure to women who not only serve as role models but offer practical advice and solutions. One faculty member described the workshop as "outstanding, as I really needed guidance in asking for things since I am the only female faculty member in my department." Finally, a comment from a graduate student captures the importance of the format that provides an opportunity " to have the interactive audience and hear what issues others have."

While the various speaker visits are typically organized as discrete events aimed at mentoring an audience or workshop group, other campus programs have been developed to foster longer-term faculty networks through mentoring. For example, the Biggio Center for the Enhancement of Teaching & Learning has introduced new Auburn faculty to the university through their New Faculty Scholars program for several years and provides presentations on professional development and group-level mentoring that extends throughout a single academic year.

To ensure that new faculty are also provided opportunities for one-on-one mentoring, including mentoring on grant-writing and publishing, the Early Career Faculty Mentoring Program was initiated in the fall of 2009. Housed in the Office of the Provost, this program conveys a strong message about the commitment of upper university administration to supporting mentoring opportunities that will enhance the success of all new faculty members in their academic careers. The program also supports the Strategic Diversity Plan goal of recruiting and retaining minority and women faculty. In February 2009, the first female Provost was hired at Auburn University, and later that year she convened a committee to plan this mentoring program. Committee members include the Associate Provost, the Director of the Biggio Center for the Enhancement of Teaching & Learning; the Diversity Faculty Mentor in the Office of Diversity and Multicultural Affairs; and the Presidential Fellow whose work in that role focused on identifying and addressing mentoring needs at Auburn University. The Assistant Provost for Women's Initiatives, who is also responsible for the activities of the *ADVANCE Auburn* Center, was asked by the provost to oversee the new mentoring program and to coordinate activities of this program with existing departmental mentoring programs and other faculty development programs on campus. Members of the committee meet regularly to discuss ways to provide support and mentoring for junior faculty.

The key elements of the program included inviting new faculty to participate and identifying mentors; coordinating activities of the new mentoring program with ongoing efforts for faculty development; creating the mentor-mentee pairs; and maintaining contact with the participants over the course of the year. A mentoring website was developed and is available via the webpage of the Office of the Provost. The program is open to new faculty who are in their first three years at Auburn. Currently, 35% of participating mentees are in STEM fields; and roughly half of these STEM mentees are female. Program mentors are Alumni Professors, an honor given to a small number of faculty members each year as an indication of excellence in teaching, research, and service to the university and larger community.

Potential mentors and mentees receive a letter from the provost explaining the purpose of the new formalized mentoring program and inviting them to participate. Interested faculty members complete a checklist of expectations for their role in the mentoring relationship. The checklists used were modified from those developed by Brainard, Harkus and St. George (*32*), which have been employed in other academic mentoring programs such as those at the University of Missouri and fellow ADVANCE recipient New Mexico State University. Participants indicate areas of professional expertise (mentors) or development needed (mentees) and amount of time they felt they could dedicate to a mentoring relationship, as well as the types of mentoring relationships in which they would be interested in participating: one-on-one mentoring, mentoring circles, peer mentoring, or having multiple mentors. The checklists are used to match mentors and mentees. A mentoring contract worksheet was provided to encourage mentoring pairs to outline the expectations and boundaries of their new relationship in writing. The pairs were asked to return these worksheets (as recommended by New Mexico State University) to the Office of the Provost because formally filing the contracts has been shown to foster accountability within a new mentoring relationship.

The Early Career Faculty Mentoring Program has helped spread awareness of the types of mentoring available to new faculty, not only through formalized mentoring pairs but also through co-sponsoring events such as the "Best Practices in Mentoring" panel with the Office of Women's Initiatives and the Provost's Office. This panel featured six tenured faculty who were recommended by their deans as outstanding mentors and are well known for their service to the institution. The main purpose of the presentation was to address mentors' interest in learning about successful mentoring strategies from their peers. Panelists addressed the need for multiple mentors to aide with different areas of development at different points in time, as well as the need for mentoring at different levels, including program, department, college, university, and career. The panel also discussed the importance of being able to give and receive critical feedback, and the necessity of longitudinal mentoring relationships.

To assess the effectiveness of this new faculty mentoring program, an electronic survey was sent to all participants in the program. Responses from the junior faculty indicated that the most helpful mentoring experiences emphasized: academic activities that will most benefit my future career; writing for publication; developing and funding research; getting resources to support professional development; working with the department chair and with senior faculty in the department; networking; time management; promotion preparation; and contract renewal and tenure strategies. In addition to identifying specific topics that were most helpful, respondents also provided comments and suggestions for changes. A majority of mentors and mentees who responded indicated that the experience was worthwhile, and all responding mentees indicated that they intended to continue and would recommend the program to other faculty. One mentee remarked that "it was helpful not to have to find my own mentor," while another noted that "the fact that such a program exists made me feel that the university recognizes that it can be overwhelming to be a new faculty and cares to make the first year more pleasant." Of particular relevance for faculty in STEM is this observation: "As a young scientist, [my mentor] has taught me how to deal with adversarial senior people in the department, and also how to develop graduate students and postdocs, and get those most interested in the lab to commit to the program." Another mentee noted that the "insight and guidance of senior mentors is extremely valuable." Mentors too indicated that they had benefitted from their participation and valued opportunities to "provide insight to assist early career faculty," "interactions with colleagues and administrators," "the opportunity to meet young faculty," and to take part in "deliberate discussions about what is important for career." Mentors noted that the overall strengths of the program included access to major leaders in the university, professional contacts between junior and senior faculty, purposeful and scheduled mentor-mentee interactions, and demonstrated administrative support for developing young faculty.

Interactive Theatre as a Small Wins Approach to Mentoring

ADVANCE Auburn has incorporated the use of theatre techniques as a way to illustrate departmental climate issues and other barriers to the advancement of women faculty, an approach used by a limited number of ADVANCE institutions as a means of drawing attention to the issues faced by female faculty. One of our team members, a professor in the Department of Theatre, combined the interactive theatre techniques pioneered by Boal (*33*) and the gender schema concept of Valian (*34*) to script two original theatrical pieces. These scripts were designed to educate departmental and university administrators on the unique impediments to success encountered by female faculty, particularly those within the STEM disciplines. Semi-structured interviews with female faculty and administrators across the STEM disciplines at Auburn provided the basis for the scripts. The interactive scripts were then performed at two annual workshops organized by *ADVANCE Auburn*. At the request of the provost, the first script was also presented at the monthly meeting of department heads and chairs. Subsequently, all heads and chairs were invited to attend the second script performance at the next year's workshop.

While both scripts explored milestones within a female STEM faculty member's career, each emphasized a different critical stage of her career. The first piece, entitled "The Third Year Review," depicts a meeting between a male department chair and a pre-tenured female faculty member and their discussion surrounding the results of the faculty member's third-year review and progress toward tenure and promotion. The exchange between the faculty member and her department chair reveals the challenges faced by both participants with respect to the communication of clear expectations for tenure and promotion and how those ambiguities prevent proper mentoring of the faculty member. In the second script, "Beyond Tenure," the scenario illustrates an informal interaction between two male and one female senior faculty member immediately following the successful tenure and promotion vote of three junior faculty, two female and one male. The dialogue explores not only the senior faculty opinions concerning the likelihood that each of these newly tenured faculty will ultimately join their ranks as full professors, but also draws attention to gender schemas that are employed by both men and women in evaluating others, and the ways that these schemas contribute to the "tiny cuts" that impede the advancement of women faculty.

The interactive theater pieces draw upon the ideas of Forum Theatre, a theoretical perspective organized around the principles of Theatre of the Oppressed which advocates for dialogue as a teaching and learning tool. The goal of writing and presenting the scenarios was to promote reflection and discussion among the audience and performers thus inspiring possible solutions to the problems faced by the characters, including effective mentoring techniques for administrators and senior colleagues and self-advocacy tools for female faculty.

Upon completion of the scenes, audience members were divided into groups to discuss questions designed to: 1) assess instinctive reactions; 2) generate reflective responses and group discussion; and 3) identify specific challenges and effective solutions within the mentoring process. By being both observers and participants, members of the audience were able to debate a range of "best practices" for mentoring and advocacy that could be implemented within their own administrative units.

The interactive theatre pieces also meet the criteria of a small win in retaining female faculty in STEM disciplines at Auburn University. Both pieces seek to create change at the departmental or college level by achieving an awareness of implicit bias and a buy-in from administrators and tenured faculty in those units. In addition, not only do "The Third Year Review" and "Beyond Tenure" advocate proper mentoring for female faculty, but they also represent another type of mentoring. Just as female faculty need mentoring throughout their careers, so too do administrators. Departmental administrators often move into their positions with little previous education on the mentoring of their faculty members – their own experiences are all they have guide them. Their personal experiences may vary substantially from those faced by faculty today. As the faces of faculty become increasingly diverse, so too will their mentoring needs.

New Mentoring Pathways at Auburn University

In contradiction to traditional intervention approaches to improving female representation in the academic STEM disciplines, simply increasing the number of women in the pipeline to a critical mass alone is not sufficient to ensure that those who come after them will be retained. While research by Kanter (*35*) suggests that while a strong minority presence of approximately 15 percent tends to improve the overall climate by gaining influence and self-perpetuation; higher percentages tend to result in the bifurcation of the minority group along generational lines (*36*). Often the most senior female faculty in the sciences achieved their success by following the model established by their male colleagues; however, younger female faculty members seek a path to academic success that allows for a balance between home and work lives. There is a need for a new generation of female STEM faculty who have successfully navigated work-family paths through academe to earn senior status as full professors and administrators. Not only would this help the STEM disciplines reach a critical mass of female faculty, but also provide appropriate mentors for those who follow their path to academic success.

Initiatives to mentor junior faculty represent small wins for Auburn University and other ADVANCE-supported institutions, but the *ADVANCE Auburn* cost-benefit analysis also revealed other important steps that can be taken toward institutional transformation for women in STEM disciplines. The results of the analysis suggest that the single most valuable small win that can influence the overall climate of a university for female faculty is facilitating female faculty participation in key academic committees. Membership in key academic committees, including those such as university-level tenure and promotion committees, allows female faculty to impact policy and the overall climate of the institution. Just as the number of female faculty must exceed a critical mass to reach beyond mere tokenism across the university and within the STEM disciplines, so too must the presence of females on influential committees reach a critical mass for true change to occur (*36*). Broad committee participation by female faculty would further contribute to the structural assimilation of women within the university and their ability to affect change. For women to obtain seats on these committees and represent the female perspective; however, they must often attain full professor status. As the "Beyond Tenure" theatrical piece illustrated, the mentoring of faculty, particularly female faculty, needs to continue even beyond the initial award of tenure and promotion to associate professor.

Auburn's current ADVANCE grant has done much to support existing junior faculty mentoring initiatives but has also recently directed efforts toward the support of women STEM faculty beyond tenure. Recently, *ADVANCE Auburn* hosted a workshop entitled "Post-Tenure Pathways" that emphasized the critical need to address the issues faced by mid-career female faculty in the STEM disciplines. Presentations focused on best practices for advancing women from associate to full professor status. As the work of *ADVANCE Auburn* suggests thus far, mentoring will play a key role in opening these new pathways.

Importantly, this discussion of mentoring efforts implemented at Auburn University reveals that successful mentoring is multifaceted. It must be offered in a wide range of formats and scopes so that one or more will be amenable to both mentors and mentees. Formal mentoring can occur at the departmental level or through university-wide programs, as evidenced in the Biggio Center's New Faculty Scholars Program and the Early Career Faculty Mentoring Program's one-on-one pairings. Equally important is informal mentoring by individual senior faculty with a vested interest in the professional success of their increasingly diverse junior colleagues. Informal networking activities for women faculty members provide an important source of social support and can offer opportunities to develop relationships with faculty outside of one's home department. Such relationships can provide safe opportunities to share experiences and tactics for navigating departmental and college climate concerns. Seminar series featuring noted women from other institutions showcase these scholars as role models for academic women in STEM fields. Additionally, their wealth of experience and advice are critical in identifying effective strategies for addressing challenges that academic women in STEM disciplines face. Successful faculty who have benefitted from effective mentoring can ultimately become strong mentors to others, provided they too have received guidance on how to mentor the next generation of faculty. Mentoring must become institutionalized; it must become the norm, not an anomaly in the academic career path for faculty. Support for mentoring at Auburn University spans across the university and has served to create organizational networks of faculty and administrators at all levels who recognize that the small win represented by mentoring can have significant outcomes.

References

1. Heylin, M. Evolving anatomy of the U.S. labor force. *Chem. Eng. News* **2005**, *83*(24), 17.
2. Congressional Commission on the Advancement of Women and Minorities in Science, Engineering, and Technology (CAWMSET). *Land of Plenty: Diversity as America's Competitive Edge in Science, Engineering, and Technology*; U.S. Government Printing Office: Washington, DC, 2000.
3. Rosser, S. *The Science Glass Ceiling*; Routledge: London, 2004.
4. Nelson, D. J. A National Analysis of Diversity in Science and Engineering Faculties at Research Universities, 2007. http://chem.ou.edu/-~djn/diversity/briefings/Diversity%20Report%20Final.pdf.
5. Long, J. S. *From Scarcity to Visibility: Gender Differences in the Careers of Doctoral Scientists and Engineers*; National Academies Press: Washington, DC, 2001.
6. Etzkowitz, H.; Kemelgor, C.; Uzzi, B. *Athena Unbound: The Advancement of Women in Science and Technology*; Cambridge University Press: Cambridge, MA, 2001.
7. ADVANCE: Increasing the Participation and Advancement of Women in Academic Science Engineering Careers, 2010. http://www.nsf.gov/pubs/2009/nsf0941/nsf0941.pdf.
8. Strategic Diversity Plan, Auburn University, 2005. http://www.auburn.edu/administration/specialreports/diversityplan/.
9. National Universities Rankings, 2010. U.S. News and World Report. http://colleges.usnews.rankingsandreviews.com/best-colleges/national-universities-rankings.
10. Schein, E. H. *Organizational Culture and Leadership*; Jossey-Bass: New York, 1985.
11. The Advisory Committee on Women's Leadership. *Issues Affecting Women at Auburn University*. Auburn University: Auburn, AL, 2004.
12. Strategic Questions That Must be Addressed. Strategic Planning Sessions, Auburn University: Auburn, AL, 2005. http://www.auburn.edu/administration/spc/may24session1,2,3.html.
13. Meyerson, D. E.; Fletcher, J. K.A Modest Manifesto for Shattering the Glass Ceiling. *Harvard Business Review*, January–February 2000, pp 127–136.
14. Kotter, J. P. Leading Change: Why Transformation Efforts Fail. *Harvard Business Review*, March–April 1995, p 59.
15. Lewin, K. Group Decisions and Social Change. In *Readings in Social Psychology*; Macoby, E. E., Newcomb, T. M., Harley, E. L., Eds.; Holt, Rinehart, and Winston: New York, 1958; pp 197–211.
16. Burke, W. W.*Organization Development: Principles and Practices*; Little, Brown, and Company: Boston, MA, 1982.
17. Spencer, M. H.; Winn, B. A. Evaluating the success of strategic change against Kotter's eight steps. *Plann. Higher Educ.* **2005**, *33*, 15.
18. Lord, M. Making It through the Maze. *Prism*, American Society for Engineering Education, October 2005, p 31.

19. de Janasz, S.; Sullivan, S. E. Multiple mentoring in academe: Developing the professorial network. *J. Vocational Behav.* **2004**, *64*, 263–283.
20. Boice, R. New faculty as teachers. *J. Higher Educ.* **1991**, *62*, 150–173.
21. Sorcinelli, M. D. The department chair's role in developing new faculty into teachers and scholars. *J. Higher Educ.* **2002**, *73*, 179–181.
22. Kram, K. E. *Mentoring at Work: Developmental Relationships in Organizational Life*; Scott Foresman: Glenview, IL, 1985.
23. Dean, D. J. *Getting the Most Out of Your Mentoring Relationships: A Handbook for Women in STEM*; Springer: Berlin, 2009.
24. Higgins, M. C.; Kram, K. E. Reconceptualizing mentoring at work: A developmental network perspective. *Acad. Manage. Rev.* **2001**, *26*, 264–268.
25. Parker, P.; Hall, D. T.; Kram, K. E. Peer coaching: A relational process for accelerating career learning. *Acad. Manage. Learn. Educ.* **2008**, *7*, 487–503.
26. Ragins, B. R.; Cotton, J. L. Mentor functions and outcomes: A comparison of men and women in formal and informal mentoring relationships. *J. Appl. Psychol.* **1999**, *84*, 529–550.
27. Egan, T. M.; Song, Z. Are facilitated mentoring programs beneficial? A randomized experimental field study. *J. Vocational Behav.* **2008**, *72*, 351–362.
28. Lankau, M. J.; Carlson, D. S.; Nielson, T. R. The mediating influence of role stressors in the relationship between mentoring and job attitudes. *J. Vocational Behav.* **2006**, *68*, 308–322.
29. Nielson, T. R.; Carlson, D. S.; Lankau, M. J. The supportive mentor as a means of reducing work-family conflict. *J. Vocational Behav.* **2001**, *59*, 364–381.
30. Allen, T. D.; Eby, L. T.; Poteet, M. L.; Lentz, E.; Lima, L. Career benefits associated with mentoring for protégés: A meta-analysis. *J. Appl. Psychol.* **2004**, *89*, 127–136.
31. Underhill, C. M. The effectiveness of mentoring programs in corporate settings: A meta-analytical review of the literature. *J. Vocational Behav.* **2006**, *68*, 292–307.
32. Brainard, S. G.; Harkus, D.; St. George, M. *A Curriculum for Training Mentors and Mentees in Science and Engineering*; The University of Washington: Seattle, WA, 1998.
33. Boal, A. *Theatre of the Oppressed*; Pluto Press: London, 1979.
34. Valian, V. *Why So Slow? The Advancement of Women*; MIT Press: Cambridge, MA, 1998.
35. Kanter, R. M. *Men and Women of the Corporation*; Basic Books: New York, 1977.
36. Etzkowitz, H.; Kemelgor, C.; Neuschatz, M.; Uzzi, B.; Alonzo, J. The paradox of critical mass for women in science. *Science* October **1994**, *266*, 51–54.

Institutional and Interinstitutional Initiatives

Chapter 6

Mentoring Initiatives for Two-Year College Faculty

Kerry K. Karukstis*

Department of Chemistry, Harvey Mudd College, Claremont, CA 91711
***Kerry_Karukstis@hmc.edu**

In the prologue to her chapter "Women Community College Faculty: On the Margins or in the Mainstream", author Barbara K. Townsend (Townsend, B.K. *New Directions for Community Colleges* **1995**, *23*, 39−46) wrote: "Because women community college faculty are understudied, we do not know how they perceive their position within the institution." In this chapter, eight female chemistry faculty members at two-year colleges share their perspectives on the status of women faculty on their campuses. The combination of institutional mission, high numbers of female faculty members even at all levels, and the range of internal and external professional development opportunities suggest a supportive climate that enables two-year college female faculty to prosper. The scope of formal and informal mentoring initiatives present at the campus level and in conjunction with professional societies is highlighted.

The Two-Year College Landscape

The two-year college system in the United States has grown to nearly 1200 institutions since the founding of the first such institution in 1901 (*2*). These institutions represent 34% of the nation's post-secondary institutions and exhibit a diversity of sizes, locations, and program offerings to meet the needs of the regions they serve. Both full- and part-time students attend two-year colleges with a variety of educational objectives, e.g., to receive postsecondary education preparation for transfer to 4-year institutions, to seek workforce development and skills training, to pursue noncredit programs as diverse as English as a second language or first-aid training, or to attend cultural events such as performances,

© 2010 American Chemical Society

exhibits, films and lecture-demonstrations. To offer such broad curricula at convenient times for the community, two-year college campuses use both full- and part-time faculty members with the needed education, expertise, and workplace experiences. The latest assessment of the number of faculty members at two-year institutions estimates the total at over 381,000 individuals, 28% of the total instructional faculty at degree-granting institutions (*3*).

Two-year colleges serve a substantial portion of undergraduate students in the United States including those receiving degrees in science and engineering (*4*). Despite the significant role that these campuses play in higher education, there have been few comprehensive efforts to document the faculty demographics at two-year institutions. Two recent estimates (*4*, *5*) of the numbers of two-year college chemistry faculty members suggest that there are at least 2600 full-time two-year college chemistry educators and comparable numbers of adjunct faculty members. According to the most recent statistics reported by the National Science Foundation (*6*), 55.2% of the scientists and engineers employed at two-year institutions were female, while the last survey of chemistry faculty (*5*) reported a significantly lower percentage of female faculty members at 32%. The American Chemical Society guidelines for two-year college programs call for institutions to provide mechanisms for the mentoring of instructional staff and for opportunities and funding for faculty renewal and professional development (*7*). To capture the status of faculty development and mentoring programs at a diverse set of institutions, eight chemistry faculty members were invited to share their responses to several questions on the career issues faced by women faculty and on the nature of existing mentoring programs on their campuses. These responses are summarized or presented verbatim below and provide an insightful view into the array of professional development situations for two-year college faculty. A brief description of each featured institution (Table 1) and a short biographical sketch of each respondent (Table 2) serve to introduce this chapter.

1. What are the demographics of women science faculty at your institution? Are there significant numbers of tenured or tenure-track women STEM (science, technology, engineering, mathematics) faculty? Are there senior women STEM faculty? Why or why not?

As the data in Table 3 indicate, faculty demographics vary widely among the institutions surveyed. A range of tenure situations also exist, from no tenure system to tenured or tenure-track status associated with all full-time faculty members. Respondents provided several insights into the strong numbers of senior women STEM faculty. For example, at Bucks Community College: "…our union contract governs such promotions." At Montgomery College: "There are as many senior women STEM faculty as there are men. There is nothing within the culture or job requirements that discourages women. The lifestyle at my two-year institution is well-suited for people with family commitments and for women raising small children." At Mt. Hood Community College: "…where I work is very child friendly (as well as pregnancy friendly) these days. Both males and females bring their kids (and babies) to work occasionally." At Georgia Perimeter College: "I attribute these numbers to the fact that our administration is made up of an equal or greater percentage of women to men, and for many years Georgia Perimeter College had a woman president."

Table 1. Locations and Enrollment Figures of Institutions Contributing to this Chapter

Bucks County Community College consists of two campuses in Buck County, PA located about 40 miles north and northeast of Philadelphia. With over 160 full- and part-time faculty, the two campuses serve about 11,000 students (10,000 students on the Newtown campus and 1000 students on the Upper Bucks County campus).
College of San Mateo is located at the northern corridor of Silicon Valley (CA) and situated on a 153-acre site in the San Mateo hills overlooking San Francisco Bay. The college has over 150 full-time faculty and nearly 300 part-time faculty serving an undergraduate enrollment of about 11,000 day, evening and weekend students.
Georgia Perimeter College is a multi-campus two-year college located in the suburbs of Atlanta, GA. With more than 25,000 students and over 1000 faculty, GPC is the third-largest institution in the University System of Georgia.
Harper College is a two-year community college located in Palatine, IL, situated approximately 25 miles northwest of downtown Chicago. The campus is spread across 200 acres. The institution's enrollment consists of nearly 26,000 students served by over 200 full-time and over 600 part-time faculty members predominantly drawn from the surrounding suburban communities.
Montgomery College serves serve nearly 60,000 students on three campuses located in Montgomery County, MD, just outside Washington, D.C. The faculty size is reported to be over 1500 members.
Mt. Hood Community College is the fourth largest of Oregon's 17 community colleges with an enrollment of over 31,000 students and more than 500 faculty members. The campus is located in Gresham, Oregon outside of Portland with an extension campus in Portland.
Pasadena City College is located in Pasadena, CA approximately 10 miles northeast of Los Angeles. The institution has the third largest community college enrollment in the United States of over 26,000 students.
San Jacinto College has three campuses in Houston and suburban Pasadena, TX serving over 23,000 students with more than 800 faculty (400 full-time).

Table 2. Two-Year College Chemistry Faculty Contributing to this Chapter

Bucks County Community College - Dr. Michaeleen P. Lee, Professor of Chemistry, Newton Campus
College of San Mateo – Kate Deline, Professor of Chemistry
Georgia Perimeter College- Dr. Candice McCloskey, Associate Professor of Chemistry, Dunwoody Campus, Dunwoody, GA
Harper College – Dr. Yvonne Harris, Dean of Mathematics and Science
Montgomery College - Susan Bontems, Associate Professor of Chemistry, Germantown Campus, Germantown, MD
Mt. Hood Community College – Dr. Elizabeth Cohen, Instructor of Chemistry
Pasadena City College – Dr. Kerin Huber, Professor of Chemistry
San Jacinto College - Dr. Ann Cartwright, Professor and Chair of Chemistry, San Jacinto College, Central Campus, Pasadena, TX

Table 3. Faculty Demographics (provided for 2009-2010)

Institution	*Faculty Demographics Provided by Respondents*
Bucks County Community College – Newtown, PA	• All full time professors at the college are tenure-track. • The college is working towards a 60-40 full to part time ratio as per the union contract. • There are senior women STEM faculty as such promotions are governed by the union contract.
College of San Mateo – San Mateo, CA	• Chemistry: Tenured full-time: 2 F, 1 M. Tenure-track full-time: 1 M. Adjuncts: 1 F, 2 M. • Biology: Tenured full-time: 4 F, 1 M. (Several adjuncts) • Math: Tenured full-time: 1 F, 10 M. Tenure-track full-time: 1 F. (Many adjuncts) • Physics: Tenured full-time: 1 F, 2 M. (No adjuncts) • Engineering: Tenured full-time: 1 F. • There are many senior women STEM faculty – e.g., 1 chemist with over 20 years; all biologists with $\geq$ 20 years, 1 mathematician >30 years, 1 physicist > 20 years, and 1 engineer > 10 yrs.

Continued on next page.

Table 3. (Continued). Faculty Demographics (provided for 2009-2010)

Institution	Faculty Demographics Provided by Respondents
Georgia Perimeter College – Dunwoody, GA	• Of the 136 total STEM faculty, 72 (53%) are women. • There are 39 tenured women STEM faculty of which 11 are full professors. • The diverse women faculty also includes 17 tenure-track professors.
Harper College – Palatine, IL	• 43 full-time STEM faculty (26 M, 17 F) teaching in 6 departments: Biology, Chemistry, Physical Sciences, Engineering, Mathematics, and Computer Science. • Full-time tenured STEM faculty: 19 M, 15 F. • 12 of the women faculty have been at Harper 10 years or more.
Montgomery College – Germantown, MD	• No tenure system exists - six-year contracts are the maximum. • Full-time chemistry faculty include 9 women and 10 men.
Mt. Hood Community College – Gresham, OR	• Mathematics is 75 % women (8 out of 10) with at least 2 have been at the institution over 7 years. • All 3 engineers are male. • In life sciences, 5 F, 7 M. • In physical sciences, 5 F, 5 M (Chemistry: 3 F, 2 M) Both physical sciences and life sciences have at least one female faculty member that has been at the campus longer than 10 years.
Pasadena City College – Pasadena, CA	• Natural Sciences Division (biology, physics, chemistry, astronomy, and geology): women comprise 66% of the tenured and tenure-track faculty and 33% of the adjunct (part time, hourly) faculty. • Many female senior faculty.
San Jacinto College – Pasadena, TX	• No tenure system. • In the Department of Science, 20 full-time faculty (9 F, 11 M). • Campus has recently instituted Distinguished Professor Levels I, II and III with plans currently being developed for Level III. • At Level II: 4 F, 0 M.

2. What kinds of mentoring programs are available to faculty at your institution? Are there any particular programs designed for women faculty? Are the faculty interested in having formal or informal mentors? Does most of the mentoring focus on junior faculty? Are there any faculty development resources directed toward senior faculty, particularly women?

The individual responses to this question show a range of formal and informal mentoring situations at two-year colleges. Their statements are presented in full to provide a complete sense of the mentoring experience from the faculty point of view.

Bucks County Community College

"Our College has a one week 'boot camp' for all full-time new faculty. In the STEM area, all current faculty are heavily involved in mentoring any new faculty. Again, no distinction between men and women."

College of San Mateo

"There are no official mentoring programs that I am aware of. There are no programs designed for women faculty. There are unofficial mentors for the new hires in chemistry since I became part of the hiring committee. The mentoring does focus on the junior faculty. Once the faculty gets tenure they are 'on their own' so to speak. There are no faculty development resources directed towards senior faculty."

Georgia Perimeter College

"The Center for Teaching and Learning offers mentoring as well as opportunities for faculty development. There are monies available for travel to conferences, videos on best practices, and opportunities to give or attend talks at the twice-yearly faculty development days. There are no programs geared especially toward women. The mentoring at Georgia Perimeter College is as formal or informal as one desires. Adjunct faculty in the STEM fields undergo a formal mentoring program and are assigned to a full-time mentor. Junior faculty are assigned to a tenured faculty mentor, and they can look to the department chairs for mentoring as well. My department chair offered much valuable advice and direction when I began teaching here."

Harper College

"We do have ... an 'established' informal relationship between faculty that recognizes the need of new faculty to have someone to guide them in navigating through various layers of tenure and promotion, as well as, acclimating to the institution at large. Usually, the mentor is a senior faculty member from outside of their department. The mentor is usually well established in their own department and expected to meet with the new faculty member regularly and serve as an informal liaison and/or a resource of information. The mentor is expected to help the faculty member through tenure, promotion, committee obligations, professional development and working through departmental, divisional and collegial relationships.

Do our existing mentoring relationships recognize the need for mentoring programs designed for women faculty ... no, we do not.

Do we as an institution have any faculty development resources directed toward senior faculty, particularly women ... no, we do not.

Although, an aggressive professional development strategy is lacking at our institution there have been several faculty who have informally stepped forward to emerge as leaders and excellent mentors. ... They are most often the senior faculty who have been involved in various initiatives across the college and have managed to befriend colleagues in other divisions and departments and at other institutions.... These faculty find themselves not only mentoring other faculty but administrators as well. I know because I seek them out often."

Montgomery College

"We mentor all new faculty, not just women. There is no particular program for women that I am aware of. We also do a lot of informal mentoring within chemistry. We share lecture notes, sample problems and exams and many other resources with new faculty to make their transition to teaching easier. The formal mentoring program only focuses on junior faculty. We have many professional development classes available to us through our institution; however, they are not directed particularly toward any group of people."

Mt. Hood Community College

"There are no mentoring programs specifically designed to mentor female faculty members, and until relatively recently there was no mentoring program for pre-tenured faculty. When I started 9 years ago, there was almost no mentoring in place at all. During the three years that it took for a junior faculty member to earn tenure, mentoring consisted of student evaluations and in class evaluations by fellow faculty members. Now, every junior faculty member has a mentoring committee that is there to answer questions and give advice. The committee also meets with the tenured faculty to discuss each non-tenured faculty member to see if there are any issues with that individual that may not have been brought to light in the various evaluation processes. If there is an area that a faculty member needs to improve, the committee then informs the faculty member and offers advice. Thus, when the faculty member comes up for tenure there are no surprises. This system has been working well for many years now. There is no mentoring in place for tenured faculty.

Although there is money still available for us to seek out our own faculty development, but there are few scheduled faculty development activities compared to what their use to be. There use to be faculty development workshops almost every week on a wide variety of topics ranging from improving your syllabus to reducing racial stereotyping in the classroom, but these are almost all gone because of budget cuts. Faculty now must find their own professional development activities outside of the college. By contract we must complete 14 hours of professional development activities."

Pasadena Community College

"We have official, administration-run orientation programs for new faculty for both men and women. Most of the 'real' mentoring of new faculty is done on an informal basis within each department .....There are faculty development resources, but they are equally directed to all tenured faculty. We are a public institution- so faculty development resources, such as funds for conference attendance and release time for special projects, have been cut severely during the last few years of budget struggles."

San Jacinto College

"We have a full-year of bi-weekly sessions for new faculty provided by the Office for Professional Development. There has been an informal mentoring system in places for several years, but this year it will be formalized and all faculty who wish to attain Distinguished Level III will be assigned two new faculty members to mentor for a year, with training, classroom visits, journaling, etc. Since we instituted an Office for Professional Development a few years ago, we have programs, workshops, etc., almost weekly for all faculty. They are usually scheduled for more than one time, so that everyone can attend, if desired. We have no special programs for women, although a chapter of the AAUW [American Association of University Women] is forming in our college district."

3. What are the major issues for women STEM faculty at your institution? Are these issues for all faculty or are their particular needs for women faculty that are not well-addressed? What kinds of programs or policies are in place to assist women faculty? What programs or policies would you like to see implemented?

Some of the respondents felt that there were few issues facing women faculty (or even faculty as a whole) at their institution.

Bucks Community College

"No major issues for women at this time. In the past, women came in at lower salaries but this was addressed by the union about 20 years ago and things are on par now."

Montgomery College

"I feel that all faculty are well-supported at my college. I do not feel that there are any major or minor issues for women STEM faculty at my college. ...the lifestyle of a two-year college professor fits well with the needs of younger women with families."

Mt. Hood Community College

"I think that the issues have already been addressed over the years. The issues most likely had to do with family, but over the years my college has become increasingly family friendly for both students and faculty."

Others felt that the issues, while varied, were equally challenging for both male and female faculty.

College of San Mateo

"All of us are being inundated with more and more administrative work which is becoming very draining. Due to the budget cuts and AB 1725 "shared governance" [California Assembly Bill 1725 ensuring faculty, staff, and students the right to participate effectively in district and college governance] we have countless extra hours for non-teaching stuff. It gets harder as the number of full time faculty shrinks and the "busy work" increases."

Georgia Perimeter College

"The major issues for all STEM faculty include pay, merit raises, workload, the adjunct-to-fulltime ratio, and program assessment. "

Pasadena Community College

"The issues we face cut equally across gender lines. Our main difficulties are a shrinking budget, not enough class sections to handle the student demand, inadequate and obsolete tech support, and aging, poorly maintained buildings."

"The problems are the same for all faculty, men and women. How to engage our students, meet their needs and help them be successful."

One respondent, Yvonne Harris of Harper College, described several broader issues that she feels should be of concern to women STEM faculty in general.

Harper College

"Moving through the processes of tenure and promotion is not a major issue for women STEM faculty at Harper. This is not a surprise since, except for the President, most of the upper administrators are women. All in all, Harper is a very good place for women faculty to be. ....."

What programs and policies would I like to see implemented at this institution? I want to see programs and policies that serve to remind women that, despite the appearances of our times, we continue to struggle for equal recognition and support and that these things are absolutely essential to having a successful mentoring program for women STEM faculty. They must serve to educate and bring awareness to our historical struggles. They must serve to enlighten our women students who seek careers in STEM related fields so they not forget who they are. They must serve to bind us together as women; as compatriots who have the understanding that we are part of something larger than ourselves. It is from here that mentoring emerges and where we begin to build the architecture of mentoring women in STEM. It is from having a 'sense of self' that having a 'sense of ourselves' become manifest in the kind of relationship that helps us guide other women."

4. How do you think the faculty development situation differs for women faculty at two-year colleges compared to women faculty at other types of institutions?

A number of the respondents indicated that they had experience teaching at a variety of types of institutions and were able to address this question with first-hand experience. Dr. Candice McCloskey at Georgia Perimeter College remarked that the experience at institutions where faculty members are expected to develop their careers based on their research programs could be a "highly individualized" and "fairly lonely" venture. "At Georgia Perimeter College, however, I am expected to further my development as an educator. And I have found that the conferences that are geared to this tend to promote and foster a feeling of community. The friendly atmosphere at these conferences makes meeting and talking to new colleagues easy, enjoyable and meaningful."

Dr. Michaeleen Lee at Bucks County Community College felt "that women are much better off at two-year schools probably due to the fact that so many of us are unionized and also that we do not work under a 'publish or perish' environment. We do have to do 'other scholarly' activities but most of these activities do not absolutely need to be done at the 'office'." Professor Susan

Bontems at Montgomery College also remarked on the availability of flexibility to enhance faculty satisfaction: "At my two-year college, women seem to thrive. They have time to be successful at work and happy at home. This comes predominantly from the flexibility of the hours. Grading and preparation can usually be done at home or at a time most convenient to the faculty member. We are not required to be at our desks from nine to five and this makes the work much more productive and fun. I work hard to get my job done and then still have time to enjoy life. It has been an excellent fit for me."

5. Is there any way that professional societies or funding agencies could help to provide the faculty development resources that women STEM faculty at two-year colleges might need? What mentoring or career development resources or activities does ACS provide through the Two Year College Chemistry Consortium (2YC₃) within the Division of Chemical Education?

Responders praised the usefulness of $2YC_3$ conferences for two-year faculty in terms of providing both professional development as educators and opportunities to serve in leadership roles. In fact, several of those surveyed indicated that their own professional development had been enhanced by election to an officer position in the Division of Chemical Education (CHED) or on the CHED's Committee on Chemistry in the Two-Year College which serves as the board of $2YC_3$.

Professor Candice McCloskey of Georgia Perimeter College offered some specific recommendations: "ACS and funding agencies can help by funding travel to conferences, by funding testing research for program assessment purposes, and by promoting the societies geared to two-year colleges, such as $2YC_3$. ACS, especially DivCHED, should bolster the presence of this relatively unknown part of ACS." Professor Yvonne Harris of Harper College suggested that the faculty themselves must take the initiative to build professional development resources and that "what ACS and/or 2YC3 can provide is the 'spark' for motivation." She offered several crucial questions for the two-year community to address:

- "How do we organize and 'knit' together women across different STEM disciplines within an institution such that they are working together in an interdisciplinary way to support and mentor each other and women students who are interested in pursuing careers in science, mathematics, advanced technologies and engineering?
- How do we promote the institutionalization of such an organization such that they are supported by the college's strategic plan?
- How do we empower these women and ensure monetary support?
- How do we build a program with these women across institutions?
- How do we provide and encourage growth and sustainability?
- How do we 'seed', expand and weave this type of mentoring for women STEM faculty into other regional and organizations such as ACS [American Chemical Society], AWIS [Association for Women in Science], AAAS [American Association for the Advancement of Science], etc. such that we have a united voice that is heard nationally?"

Indeed, these are significant questions for all types of higher education institutions to address to insure the participation, promotion, leadership, and visibility of all women STEM faculty.

Future Directions

With projections for increasing enrollment at two-year colleges in the future and the associated need for expanding faculties, consideration of two-year college mentoring and faculty development programs is timely. Recognizing that more deliberate efforts might be necessary, some interesting models for mentoring programs exist beyond individual campuses.

For example, the American Association of Physics Teachers (AAPT) conducts a 14-month experience designed specifically for two-year college physics faculty in their first five years of teaching. Known as the "Two-Year College New Faculty Experience" (http://www.aapt.org/Conferences/newfaculty/tyc.cfm), the program consists of attendance at two summer AAPT national conferences and online discussion sessions during the intervening fourteen months. The experience is led by experienced two-year college faculty and is designed to equip new faculty members with knowledge of three major active-learning pedagogical initiatives in introductory physics that have improved student comprehension and that have been successfully implemented at two-year colleges across the country. While the activities are clearly geared toward communicating pedagogical innovations, participants are likely to begin to build a support community of both fellow new faculty and experienced mentors. This initiative, while organized around a professional society's annual conference, is supported by the National Science Foundation and represents the kind of partnership that will facilitate the professional development of cohorts of community college faculty.

Another partnership involving the National Council of Instructional Administrators (NCIA, www.nciaonline.org) and the Council on Undergraduate Research (CUR, www.cur.org) conducts workshops for community college faculty interested in incorporating undergraduate research into the curriculum. This initiative, also supported by the National Science Foundation, aims to provide a workshop curriculum to implement undergraduate research programs tailored to the needs of community colleges and to develop an undergraduate research mentoring network of community college faculty. Once again, while the focus of the workshop is pedagogical, the experience generates mentoring relationships that can continue beyond the workshop offering.

The American Association of Community Colleges (www.aacc.nche.edu) notes that "Community colleges are in the midst of a transition brought about by the numerous retirements of administrators and faculty members." As such, this is an opportune time to consider new mentoring initiatives and faculty development programs as a whole. Many of the successful career development examples shared here as well as the thoughtful suggestions offered should prove invaluable as institutions both hire new faculty to meet future demands and design the supporting infrastructure that will enable them to thrive.

References

1. Townsend, B. K. *New Directions for Community Colleges* **1995**, *23*, 39–46.
2. Digest of Education Statistics: 2009, Table 266. Degree-Granting Institutions and Branches, by Type and Control of Institution and State or Jurisdiction: 2008-2009. United States Department of Education. http://nces.ed.gov/programs/digest/d09/index.asp (accessed October 2010).
3. Digest of Education Statistics: 2009, Table 249. Number of Instructional Faculty in Degree-Granting Institutions, by Employment Status, Sex, Control, and Type of Institution: Selected Years, Fall 1970 through Fall 2007. United States Department of Education. http://nces.ed.gov/programs/digest/d09/index.asp (accessed October 2010).
4. Ungar, H.; Brown, D. *J. Chem. Educ.* **2010**, *87*, 572–574.
5. Ryan, M. A.; Neuschatz, M.; Wesemann, J.; Boese, J. *J. Chem. Educ.* **2003**, *80*, 129–131.
6. Women, Minorities, and Persons with Disabilities in Science and Engineering; NSF 09-305; 2009. Division of Science Resources Statistics, National Science Foundation. http://www.nsf.gov/statistics/wmpd/ (accessed October 2010).
7. ACS Guidelines for Chemistry in Two-Year College Programs, 2009. ACS Committee on Education, American Chemical Society. http://portal.acs.org/portal/PublicWebSite/education/policies/twoyearcollege/CNBP_025786 (accessed October 2010).

Chapter 7

Support from Academe: Identifying Departmental and Institutional Resources, Policies, and Infrastructure to Support Senior Women STEM Faculty

Ruth Beeston,[1] Jill Granger,[2] Darlene Loprete,[3] Leslie Lyons,[*,4] and Carol Ann Miderski[5]

[1]Department of Chemistry, Davidson College, Davidson, NC 28035
[2]Department of Chemistry, Sweet Briar College, Sweet Briar, VA 24595
[3]Department of Chemistry, Rhodes College, Memphis, TN 38112
[4]Department of Chemistry, Grinnell College, Grinnell, IA 50112
[5]Department of Chemistry, Catawba College, Salisbury, NC 28144
*lyons@grinnell.edu

In this paper we discuss institutional and departmental mechanisms which support women faculty in chemistry at liberal arts colleges at all stages of their career, from the pre-tenure years through retirement. For senior women faculty who will be in their careers for the longest time span, we recommend six policies which include sabbatical leaves, travel support, faculty development efforts, family leave policies (including elder care), medical leave policies, and phased retirement options. Shared/split academic positions are also discussed as a mechanism for the academy to bring more women into academic positions and promote a better work-life balance. Flexible implementation of policies can provide the broadest participation of women in these career advancement mechanisms.

© 2010 American Chemical Society

Introduction

The presence and success of women in academe, especially in the sciences where women have been and continue to be under-represented nationally, can be constructed and supported by the academic infrastructure of institutions and departments. In contrast, the absence of women in programs can be directly tied to the absence of effort and/or lack of support for women in particular locales. The National Academies' comprehensive study of women in science and engineering is a treasure trove of information on the problems and a call for change (*1*). As well, the glacial pace of change in the academy modeled by Marschke et al. will cause gender equity in the future to be just as elusive due to demographic inertia (*2*). In this paper we will illustrate mechanisms that have led to the success of women in liberal arts chemistry departments.

We will use the typical academic career path of hiring, tenure, and post tenure professional development as our roadmap. As the focus of our NSF-Advance-Paid project is on senior women faculty, we will go into more depth on the post tenure professional development mechanisms. But, we won't ignore how we got to be senior women faculty at our institutions and reflect on hiring and tenure policies briefly. Many of these mechanisms can be generalized to other areas in which women are under-represented in the work force.

Hiring

Senior women faculty at liberal arts colleges began their careers with an initial hire into an academic department, usually at the assistant professor level. Success in hiring women into broadly advertised, open positions can be most successful if the department or institution makes a serious effort to attract female applicants. This paper will not argue the importance of women's representation in departments, leaving that topic to other published work.

The efforts to attract female applicants need to go beyond the legally required equal opportunity language. Institutions that welcome diverse applicants have augmented language that encourages women and minorities to apply. For example, every ad at Rhodes College has the following language at the end of the ad:

> Founded in 1848, Rhodes College is a highly selective, private, residential, undergraduate college, located inMemphis, Tennessee. We aspire to graduate students with a lifelong passion for learning, a compassion for others, and the ability to translate academic study and personal concern into effective leadership and action in theircommunities and the world. We encourage applications from candidates interested in helping us achieve this vision. Memphis has a metropolitan population of over one million and is the nation's 18th largest metropolitan area. The city provides multiple opportunities for research and for cultural and recreational activities. Read more about Memphis at http://www.rhodes.edu/about/369.asp We are an equal opportunity employer committed to diversity in the workforce [http://www.rhodes.edu/collegehandbook/10309.asp].

The goal of the ad language is to attract a wide variety of qualified applicants that share the Rhodes College mission. They wish to hire people that will embrace different cultural perspectives, contribute to a liberal arts college environment and educate students in a way that fosters respect and intellectual stimulation. The ad language accomplishes this, in part, by using the college handbook link that describes their commitment to diversity which is also tied to their multicultural affairs webpage. An applicant can read about the kinds of diversity supported by the institution (for example, gay-straight alliance, black student association etc) and feel confident that the College supports women and minorities.

Institutions with diversity officers (also affirmative action officers) who participate and impact searches can also increase the opportunities to hire a woman into an open position. Real participation means inclusion in the formulation of the ad, in departmental discussions of search candidates, and a voice in the decision on which candidates to bring for a campus interview. Institutions with marginalized diversity officers are unlikely to improve the number of women on their faculty. For example, the diversity officers at Rhodes College meet with each candidate during their campus visit to articulate the College's commitment to diversifying the faculty. The American Chemical Society's recent inclusion of the Academic Employment Initiative at fall national meetings provides all employers a forum to connect advanced graduate students and post-doctoral students who will be seeking jobs (and are disproportionately from under-represented groups) with institutional representatives who are searching for faculty candidates. Academic institutions need to expend the resources to send a representative to this meeting because it is an excellent venue to recruit diverse faculty to an open position.

Upon hiring, start-up funds, salary, laboratory space, and initial teaching assignments can set the stage for success or failure of a new faculty person. New women faculty typically benefit from a helpful advocate to negotiate the best possible starting position in a department. Valuable assistance can also come from off-campus organizations such as CUR (the Council on Undergraduate Research), and the YCC (the Younger Chemists' Committee of the American Chemical Society) who offer career planning advice to job candidates independent of any particular institution. Lack of support or resources to conduct research or develop courses can undermine any career at its start.

Pre-Tenure Years

During pre-tenure years women frequently need to balance their burgeoning careers with family responsibilities. At the typical age of new faculty, family demands often involve young children. Institutional policies for pregnancy and family leave make this balance more successful. Not surprisingly, more generous policies lead to a higher level of success for women faculty. Resources such as an on-campus day care center provide significant support, allowing faculty with families to be both parents and successful employees. Policies that do not consume additional resources but can support faculty with family responsibilities include:

a) willingness to allow flexibility in scheduling (such as scheduling courses so as to not conflict with daycare/school start and end times),

b) informational assistance on local daycare availability,

c) willingness to schedule meetings at times that don't conflict with family responsibilities, and

d) tolerance of arrangements that include the occasional presence of children in academic buildings.

Formal mentoring, typically of a vertical nature, can benefit young faculty. The most common form of these arrangements pairs an experienced faculty member with a novice faculty member. The experienced faculty member can advise the novice on decisions regarding teaching, research, service, and the peculiarities of the institution. Such advice can be much more efficient than a trial and error approach to such decisions. Horizontal mentoring of young faculty can also be beneficial. Davidson College has an organized group of young faculty who meet regularly for mutual support and discussion of issues. Grinnell College also has a more socially construed group to serve the same purposes. Tenure brings the necessary (if unfortunate) eviction from this coterie.

All young faculty benefit from a progress review or reviews prior to the tenure decision. This is regularized at several institutions in the form of an interim review after one year of teaching and a complete review after three or four years of teaching. The interim review focuses on teaching while the complete review includes all the components of a tenure review except the external review of scholarship. An important goal of the reviews is to give young faculty feedback on areas which need improvement in order to achieve tenure. This is an investment of internal resources (and senior faculty time) but the tenure success rate is much improved. When the pre-tenure reviews are conducted from the developmental point of view instead of a purely critical point of view, the young faculty member also receives important suggestions for their own career development. For both of these reviews at Grinnell, the candidate receives a carefully constructed written progress evaluation that details specific objectives for improving job performance if necessary.

With many liberal arts colleges including a grants component as part of scholarship, all faculty and especially pre-tenure faculty can benefit from an on-campus grants officer. This person can keep faculty apprised of grant programs for which they might be eligible, assist with grant preparation (such as NSF Fastlane submissions), and keep faculty members on a schedule to meet the grant deadlines. It is important that young faculty see examples of successful proposals from their department or institution and have a resource/mentor on campus who can critique proposals and make suggestions for revisions and resubmissions of proposals.

Pre-tenure sabbatical opportunities greatly assist young faculty in meeting the scholarship demands of tenure. Typically in chemistry departments at liberal arts colleges, faculty work with undergraduate students on research projects. The presentation and publication of the resulting work at the professional level depends solely upon the expertise of the faculty member. A pre-tenure sabbatical can be essential support from the institution so that manuscripts and presentations can be prepared with some release from regular teaching responsibilities. Grinnell College provides all pre-tenure faculty with a one semester release from teaching in the fourth or fifth year of service and a competitive program to support two or three pre-tenure faculty with full year releases.

Post-Tenure Years: Senior Women Faculty

Our NSF-PAID-Advance project has focused on senior women faculty at liberal arts colleges. Appropriately so, this paper will discuss the institutional and departmental role in promoting the success of women who have achieved tenure at their institutions. In a faculty career of 30 years, almost 70% of a career is spent in this phase, so both individuals and institutions can benefit from enhancing senior faculty success.

Institutional and Departmental Support

Professional development support is important for senior faculty (*3*). Sabbaticals are essential for continued research productivity and updating of courses to continue to reflect current advances in science. Flexible sabbatical policies which support faculty at their home institution or off-campus, support faculty work on broad ranges of projects (both research or teaching projects), and flexible timing (calendar year or academic year, full year or 1 semester) can offer senior women as many options as possible to continue their professional growth. Replacement of faculty during a sabbatical leave is essential because otherwise, a department or its members are disadvantaged when one of their colleagues takes a sabbatical. If courses will be lost in a department or colleagues will have to take on an overload for the faculty member on leave, the pressures inside a department may result in less frequent sabbatical experiences.

Travel funds also are essential for senior faculty to engage with the other experts in their subdiscipline. Broadly managed programs which support conference attendance (both domestic and international), workshop participation, grant review panels, scientific society service (especially at the national level), and site visits to make specific changes to an on-campus program or facility are all examples of the types of travel support senior women faculty will need over this phase of their career.

Over the twenty or so years as post-tenure faculty, senior faculty will need to update, retool, or shift their research and teaching agendas. Faculty development support from departments and institutions are essential for success in continuing to be a productive scholar (*4*). Sabbaticals are one form of tangible support. Support for collaborative projects is another measure of career enhancement. Liberal arts chemistry departments have typically placed more value on single investigator projects but as the research topics in chemistry become more interdisciplinary (and complex), collaborative projects are becoming more common. Valuing collaborative outcomes is a significant paradigm shift which will become more important as interdisciplinarity becomes a more integral part of the curriculum at liberal arts colleges.

Tangible institutional and departmental support for new smaller initiatives can come in the form of time or money. Release time from a portion of teaching responsibilities, such as a one lab reduction in course load could support a faculty member to develop a new experimental technique or a course module. Monetary support can fund travel for a site visit or supplies to implement a new idea on the home campus.

Chemistry is an equipment intensive discipline. Updating and maintaining the instrumentation that drives continued success in research requires institutional and departmental commitment. As much of this equipment must be acquired by grant success, support from the institution is essential. Types of support include cost sharing the purchase of equipment, and funding maintenance contracts or training programs for campus users. The institution can support faculty grant success with administrative support such as a grants officer and also promote and reward institution wide proposals to foundations such as the Howard Hughes Medical Institute science education grants.

Senior faculty continue to need support to balance work and life issues. A broad family leave policy which extends to aging parent needs can support faculty who at some point in this phase of their career will face care needs of their elder generation. It is also possible that an illness or injury will require leave time at some point. Medical leave policies which include the hiring of short term replacements are important so that faculty can regain their health and their colleagues are not overburdened by extra responsibilities. A schedule of graduated work responsibilities can aid a faculty member in healing and partial return to their regular duties until the full load can be resumed. Transparent policies on unpaid leaves or fractional appointments can assist faculty in returning to their full duties gradually.

As senior women faculty approach retirement, flexible retirement options such as phased retirement plans can maintain involvement in a department with reduced duties. In departments with very few women, the retirement of even one senior woman can dramatically change the gender balance of the faculty. Grinnell College has a program of phased retirement called senior faculty status (SFS) in which faculty who are 61 years of age can renegotiate their duties with compensation at half of their salary. The range of duties is highly personal with some faculty pursuing only research or teaching while others blend service, research and a small amount of teaching arriving at a half time level of duties to preserve benefits. Departments can typically a hire a full time replacement for a faculty member who moves to SFS so that no courses are lost to a department. Since salaries are so different for a new hire and a very senior faculty member, the half a senior faculty salary frequently is comparable to a new hire's salary. At Grinnell, faculty can remain in SFS status for five years and renegotiate the terms of their half duties annually.

Emerita faculty can continue to be professionally active with the support of an on-campus office, computer support, and laboratory space/supplies. These women continue to serve as role models for faculty and students at the institution and regionally or nationally. Well known research scientists such as Larry Dahl (U. Wisconsin-Madison), John Roberts (California Institute of Technology) and I. M. Kolthoff (University of Minnesota) continued their teaching and experimental work many years beyond their 'retirement'. At Grinnell College, Ken Christiansen from the biology department has continued his research work with students with an active publication record for twenty-five years beyond the end of his teaching career.

The theme of all of the resources, policies, and infrastructure mechanisms discussed above is **flexibility**. Women's support needs frequently don't fit the institutional rubrics which have served men at liberal arts colleges for decades (and were frequently developed by men). Refining of institutional policies to offer more breadth of implementation can expand them to support women. Life and career balance issues for women can mean that the year in which one is eligible for a sabbatical may coincide with family circumstances that would cause a sabbatical away from home to be too disruptive. A sabbatical which could also be held at the home institution is a small example (though a policy shift for some institutions) which would support a faculty member's scholarly activities and balance a family need. Institutional and departmental agility can benefit women by helping them to work around rigid, out-dated programs that would serve to exclude them from participation. The willingness of administrators to discuss alternate implementation strategies and revise programs can be more broadly supportive of all faculty.

In summary, valuable institutional mechanisms that support senior faculty include:

1) sabbatical leaves,
2) travel support,
3) faculty development efforts, such as:

    a) workshop participation,
    b) recognition of collaborative projects,
    c) partial release time or modest supply funds for small projects, and
    d) new equipment acquisitions.

4) family leave policies which include support for aging family members,
5) medical leave policies which include replacement of faculty on leave, and
6) phased retirement options.

Women Supporting Women

Women can participate in activities that support each other with no or modest support from their institutions and departments. Over some organizational range (from departments up to entire institutions), there are now a critical mass of women to draw together for mutual support. Regular gatherings such as the Friday afternoon wine and snack group, Beatrice, at Rhodes College can offer regular fellowship and friendship. Davidson College has an email distribution list for women faculty. A local chapter of Iota Sigma Pi can connect women in one institution to others in the area such as the new chapter at Sweet Briar College. Carol Ann Miderski at Catawba College has established a Women's Resource

Network in North Carolina to draw women within a 100 mile radius to share information and to gather occasionally for networking and support. Grinnell College has an organization called Scholarly Women's Achievement Groups (SWAG) composed of small groups of 3-5 women faculty (both vertical and horizontal groups) that support women faculty in keeping their scholarly goals moving forward. SWAG has received some administrative and food support from the institution but the bulk of the small group meetings require only time from the women who participate. The small groups typically meet 4-6 times over an academic year.

These example activities range in their institutional support levels. Some require no monetary support from the institution (in the case of Beatrice at Rhodes College) to modest support (administration of an email list, institutional recognition of Iota Sigma Pi, administrative support and a small food budget in the case of SWAG at Grinnell College which is supported externally by the Mellon Foundation). These kinds of activities support women faculty with minimal institutional investment.

The Shared/Split Contract

A shared employment contract or a split position can provide a mechanism for career and life balance in any work environment. Typically these employment contracts hire two people (usually a partnered couple) into one faculty position at an institution. Linguistically, liberal arts colleges call these arrangements shared contracts while universities label them split positions. Several liberal arts colleges offer this employment option to faculty including Grinnell College (anthropology, biology, chemistry, mathematics and statistics), Albion College (chemistry), Alma College (chemistry), and Franklin and Marshall College (geosciences), while Calvin College (chemistry), Knox College (physics), and Whitman College (English) had faculty in these positions (5). As early as 1985 a study of twelve couples identified the lifestyle advantages of job sharing (6). The American Chemical Society Division of Chemical Education regularly sponsors a symposium on careers and in 2006 included dual career chemistry couples in both academia and industry (7).

At Grinnell, the terms of employment include sharing the teaching responsibilities of a regular faculty load, benefits for both persons, separate offices, and individual performance evaluations. A nice (and humorous) essay on the shared contract at Grinnell was published in 1999 (8). Historically, shared contract positions date from 1964 at Grinnell and, in part, were driven by diversity efforts in the 1980's and 1990's to hire women faculty. Lyons began her career at Grinnell (as the first woman in a tenure track position in a physical science department) in a shared contract with her spouse (an arrangement that continues today). At universities, job sharing arrangements are more commonly called split positions (9) in which contracts to each individual are tied to one faculty position but the distinctions between split and shared positions are more specific to institutions. Oregon State University marine biologists Jane Lubchenco and

Bruce Menge have described their arrangement in a split position as "the sane track. "(*10*)

There are advantages for both the individuals and their institutions with shared/split positions. The individuals can live and work in the same place, both have satisfying employment, and both have more time for research, teaching and family/life choices. Institutions gain faculty diversity, stability of their faculty (less attrition, experienced faculty who can cover other leaves, and less reliance on temporary faculty), better spousal employment, and in many cases the added expertise/breadth of two research programs. On a 'green' note, the carbon footprint of shared/split positions is lower as there is less commuting and one shared home vs. the long distance, two residence arrangement of so many working couples in the United States.

Disadvantages for the partners include a single salary and the potential for exploitation. While the contract may be for one position with one salary, actual implementation of these arrangements frequently includes negotiation for additional teaching (for commensurate additional pay) and financial support for the individuals via external grant support. Over the twenty-one years Lyons and her spouse have shared a position, their teaching load has ranged between 1.2 and 2.0 FTE with an average of 1.7 FTE the past seven years. The exploitation comes from two sources: the institution and the partners. The institution frequently asks the partners for more service (two people to serve on committees) and the two faculty may spend more total time at their job than one faculty person would. The institution also faces hurdles and challenges in these positions. Revisions of hiring, tenure, and evaluation procedures to establish institutional policies for shared/split positions are significant barriers. Institutions with nepotism policies may also find shared/split positions conflict with that policy. Institutions face increased costs of both individuals in the partnership versus a one person hire. Typically, these include full benefits for both, offices for both, and research support for two research programs if that is a hiring goal.

Women scientists are disproportionately in partnered relationships with other scientists (*11*) and are a part of the increasing trend of dual career couples in the United States. Partner accommodation policies from 16 institutions are collected by Eric Jensen on his web site (*12*). The shared/split employment arrangement is one solution to the two-body problem in academia. Shared/split positions are also a mechanism for the academy to bring more women into academic positions and promote better work-life balance.

Conclusions and Concerning Trends

The current economic climate and trends in higher education raise several concerns that may impact women faculty more strongly. Budget shortfalls are now part of the common news cycle (with the protests in March, 2010 receiving the most attention). The erosion of distribution requirements in the sciences at institutions is of general concern for science literacy in America. Frequently it has been women faculty who have taught and developed courses for non-science majors. The loss of these courses in the curriculum will disproportionately impact women faculty (who will then need to retool to teach other courses or face job loss). Changing enrollments are putting pressures on institutions generally. Class sizes are being pushed up, and private institutions without strong financial aid support for students are facing declining enrollments as their costs go beyond the reach of recession strapped families. Faculty are being asked to do more with less and women faculty see themselves as more vulnerable to the uncertainties of these trends. Changes in the undergraduate chemistry curriculum adopted by the ACS-CPT (American Chemical Society Committee on Professional Training), in the MCAT exam, and medical school requirements will also impact liberal arts chemistry programs.

Gender equity issues in science and at liberal arts colleges will be with the academy for some time. Glaciers in the continental United States will likely melt before gender equity is achieved broadly in the sciences, an example of geological processes outpacing sociological ones. Facing this challenging future, the departmental and institutional mechanisms we discuss here can be a part of improving the future for all faculty at liberal arts colleges with our students receiving the greatest benefits. Mechanisms which are flexible in their implementation support individual faculty to balance their work and life needs while leading the academy into the twenty-first century.

References

1. National Academies (National Academy of Science, National Academy of Engineering and Insitute of Medicine). *Beyond Bias and Barrier*; National Academies Press: Washington, DC, 2006.
2. Marschke, R.; Laursen, S.; Nielsen, J. M.; Dunn-Rankin, P. *J. Higher. Educ.* **2007**, *78* (1), 1–26.
3. National Academies (National Academy of Science, National Academy of Engineering and Insitute of Medicine). *Beyond Bias and Barrier*; National Academies Press: Washington, DC, 2006; Chapter 5.
4. Gillespie, K. J.; Robertson, D. L.; Bergquist, W. H. *A Guide to Faculty Development*, 2nd ed.; Jossey-Bass: Hoboken, NJ, 2010.
5. At these 3 institutions, the positions are no longer shared due to a) the sudden death of Karen Muyskens at Calvin, b) the departure to two positions in industry and academia by the couple at Knox, and c) the retirement of the individuals at Whitman.
6. Mikitka, K. F.; Koblinsky, S. A. *Fam. Consum. Sci. Res. J.* **1985**, *14* (2), 195–207.
7. Rovner, S. L. *Chem. Eng. News* **2006**, *84* (19), 35–38.
8. Montgomery, M.; Powell, I. In *Starting Our Careers: A Collection of Essays and Advice on Professional Development from the Young Mathematicians' Network*; Bennet, C. D., Crannell, A., Eds.; The American Mathematical Association: Providence, RI, 1999; p 9.
9. Katterman, L. *The Scientist* **1995**, *9* (21), 16.
10. Lubchenco, J.; Menge, B. *Biosciences* **1993**, *43* (4), 243–8.

11. Dual-Science-Career-Couples. http://physics.wm.edu/~sher/dualcareer.html (accessed June 1, 2010).
12. Resources for Academic Couples. http://astro.swarthmore.edu/~jensen/couples.html (accessed June 1, 2010).

Chapter 8

Why Does Mentoring End?

Cindy Blaha,[1] Amy Bug,[*,2] Anne Cox,[3] Linda Fritz,[4]
and Barbara Whitten[5]

[1]Department of Physics and Astronomy, Carleton College,
One North College Street, Northfield, Minnesota 55057-4016
[2]Department of Physics and Astronomy, Swarthmore College,
500 College Avenue, Swarthmore, PA 19081
[3]Department of Physics, Eckerd College, 4200 54th Avenue South,
St. Petersburg, Florida 33711
[4]Department of Physics and Astronomy, Franklin and Marshall College,
P.O. Box 3003, Lancaster, PA 17604-3003
[5]Department of Physics, Colorado College, 14 East Cache La Poudre Street,
Colorado Springs, CO 80903
*abug1@swarthmore.edu

Mentoring often implies an apprentice-expert relationship,
but mentoring can and should take many forms. This paper
describes a horizontal mentoring alliance of five senior women
physics faculty from small liberal arts colleges supported
through the NSF-ADVANCE (PAID) project. After a brief
review of the literature on the value of mentoring, this paper
describes the unique challenges and demands of senior women
scientists at liberal arts institutions and the ways in which this
particular alliance helped the members successfully navigate a
variety of professional and personal issues. It highlights the way
in which the alliance was instrumental in strengthening each
member's professional research whether it was an extension
of current work or changing to a new sub-field. Through the
experiences of the alliance members, this paper argues for
sustaining and propagating similar networks and suggests some
initial steps to accomplish this.

© 2010 American Chemical Society

Introduction

An email appeared out of the blue from someone I did not know at Harvey Mudd College. It invited me to participate in an NSF-sponsored mentoring alliance for senior women faculty in physics at liberal arts colleges. "Why?" I asked the email, "Why me? Why a mentor? I am a mentor to students, junior faculty in my department, and women faculty across the sciences. Why would someone mentor me? And what is "horizontal mentoring" anyway?"

This, to varying degrees, describes the initial reaction of members of the physics of the NSF PAID-ADVANCE initiative, "Horizontal Mentoring Alliances to Enhance the Academic Careers of Senior Women Scientists at Liberal Arts Institutions" (1). This reaction was based on the standard model of mentoring exemplified in Greek mythology where the goddess Athena takes the form of "Mentor" to Telemachus, the young son of Odysseus, in the Odyssey. Athena, in the form of the older and wiser Mentor, gives advice and provides encouragement and support. As Mentor, Athena helps Telemachus find his own identity, apart from his father. As faculty, we often embody this character for our students and our younger colleagues. Hence our surprise at the email. It never occurred to us to ask: Who will mentor the "Mentor"?

As I re-read the email, I wondered, "Did someone know? Has someone seen through my façade of being an organized, efficient, successful senior faculty member? Does someone know I need mentoring, even now at this advanced point in my career?"

When we consider the subject of mentoring in the context of academic careers, our discussions often focus on the earliest career stages. Many papers and articles highlight the importance of good mentoring in graduate school and early career environments (2). Much literature focuses on advice for new faculty members (3, 4). But why would one think that the need for good mentoring ends when a faculty member achieves tenure? As faculty members, we face changing needs and expectations in our teaching and research, in our contributions to campus and professional communities and in the continual struggle to balance our personal and professional lives. Wouldn't a good network of mentors be extremely helpful in navigating an ever-changing career path? The mysterious email started us down a path where we experienced the benefits of mentoring at an advanced stage in our careers. This allowed us to eventually ask "Why Does Mentoring End?"

Mentoring

Importance of Mentoring

Why is mentoring needed? The scientific community is not, as it is sometimes naively characterized, a group of individuals struggling alone to produce work whose quality is manifest to all. Rather, it is a community of overlapping and interacting networks (5). Members of a network provide support at all stages of the scientific enterprise. As scientists, we ratify and critique each others' plans, offer encouragement when work or personal life does not go as hoped and celebrate and publicize each others' success in a way that leads to career advancement. Furthermore, in order to increase our pool of scientific talent, it is generally agreed that we must open our community to race, gender and other kinds of diversity. It stands to reason then that we must affect much of this change by leveraging a network-laced structure. We must build and maintain professional networks, and help our colleagues do the same.

The literature makes it very clear that social networks in science are powerful entities. For example, the study by Wennerås and Wold (6) shows that the likelihood of getting grants is increased by knowing someone (or knowing someone who knows someone) on the committee. Blau et al. (7) compared young women economists who had participated in a mentoring workshop to those who had not. Women who participated in just this one workshop had more publications overall, more top tier publications, and more successful grant proposals. (It was too early to tell whether the workshop had an effect on tenure.) A recent National Research Council report found that 93% of women with mentors received funding for grant proposals compared to 68% of women without a mentor (8).

Etzkowitz et al. (9) have discussed the importance of 'social capital' to a scientific career. A store of social capital resides within a network of collegial individuals and is accumulated by the exchange of valuable items like scientific information, career advice, and good recommendations. Social capital is needed if we are to maintain the other types of capital—financial, physical, and human—essential for ongoing scientific success, even when we are at advanced faculty rank. They note that "Formal positions are only a rough indicator of success, since individuals of the same rank differ widely in the strength of their networks and their access to scientists with relevant knowledge for possible collaboration." (Reference (9), p. 124). Etzkowitz et al. maintain that the lack of adequate social capital provides a framework for analyzing the differences in "the success of men and women in a social context in which productivity is based on managing interdependence with others." (Reference (9), p. 118).

Traditional Mentoring and Women Scientists

A scientific career usually begins in graduate school, but of course, grad school cannot teach a scientist all she/he needs to know. It is primarily focused on training research scientists and providing the knowledge and skills necessary for doing research in a specific subfield. Grad students are coached as scientific apprentices and a thesis and post-doctoral work are their "journeyman's" projects, a sign that they are ready to practice their craft in their professional community. But after leaving grad school (and perhaps a postdoctoral position as well), career paths diverge and evolve. The trained research scientist may need to know how to be a good teacher or staff scientist, department chair or lab manager, campus administrator or program director. This suggests that career mentoring may be valuable, not just at the outset, but throughout an entire career.

What types of mentoring are useful in a scientific career? Are effective mentoring methods the same for women and men? In the physical sciences, men outnumber women at all academic ranks and the gender disparity is most pronounced at senior academic levels. (See the section below on "Our Demographics".) It is important that all scientists, regardless of gender, be able to serve as effective mentors. In order to examine the role of gender in mentoring relationships, one of us (10) has done interviews with a number of women physical scientists. Their mentoring experiences, while not presumed to be comprehensively representative, illustrate a variety of mentoring situations. (Pseudonyms will be used in the following descriptions.)

Marilyn, an African American physicist, had had an elite education and understood about networks. But she couldn't make the existing networks work for her. As a student, she and other Black students were not introduced to recruiters and other important speakers. She did favors for others but couldn't call on them for return favors.

Christine, a Native American geologist, has been spectacularly well mentored in some ways. A famous geologist called her up when she was a graduate student because they were working on the same type of rock. He took her all around the area, sharing his deep knowledge. He phoned her to ask how she was doing; when she was broke, he gave her money to pay her bills and buy her kids Christmas presents. But more than one professor at her own institution subjected her to discrimination and poor advice. She ended up having to change schools because of this bad treatment by professors and advisors.

Jane, a White mathematician, worked with Richard, another "different" scientist (he is blind). He was her Ph.D. advisor, and Jane has continued to work with him her entire career. They both consider the other to be their best collaborator. They feel lucky to have found each other, and to be able to work together consistently. Yet Jane feels that her career is incomplete because she has never become independent of her advisor.

Dolores, a Hispanic physicist, educated at an elite institution, had a well-known advisor who told her to "You have to run as far away from me, as far away from guys as possible". The advice was perhaps well-meant— intended to force Dolores to craft her own professional identity (in the way that Jane did not, to her perceived detriment)— but taking that advice put Dolores at a disadvantage. In her postdoctoral years, she was deprived of an important source of advice and influence. She had to compete with others who kept closer to their mentors and could be advised, recommended for speaking opportunities, and helped with writing grants and interacting with granting agencies.

These examples illustrate that women scientists can certainly be mentored effectively by men. Sometimes such relationships turn out quite well. But there are risks. The advice to Dolores to separate herself from her advisor was well-meant, but ultimately injurious to her career. For various reasons, some male mentors don't actively mentor women in the comprehensive way that they would mentor another man. Nolan et al. (11) surveyed early career chemists about their mentoring experiences. Women tended to have equal academic mentoring with men (e.g. research meetings with advisors), but less professional mentoring (career-building opportunities, advice on career choices). They found, however, that when women were mentored by women faculty, these discrepancies disappeared.

Mentoring in Academia: Life-Long Mentoring

When is the right time to "leave the nest", and pursue a career in science, sans mentor? In her talk "Crossing the Bridge" at the Feb. 2010 American Physical Society meeting, Fisk University student Erica Morgan described the various mentors she has encountered as she progresses toward a Ph.D. in physics. When asked at what stage in her career she would no longer need to benefit from mentoring, she was very definitive. "Never!" she said.

The premise of this white paper is that a scientist should *never* find herself/ himself without a mentor. Moreover, a scientist needs not just one, but a network of people able to serve as mentors. The types of issues that scientists face will change over the course of an academic career. Networks of mentors can provide ongoing support and encouragement throughout the various stages.

Early Years

As we've mentioned above, mentoring efforts in academia are typically aimed at faculty just beginning their careers. New faculty need help in order to learn how to teach, establish a research program and write successful proposals to supplement start-up funds and support ongoing research. Without missing a beat, they must go from being grad students (or postdocs) to serving as advisors to their own students. For some women, this transition is complicated by domestic issues like managing a two-career family and raising young children.

In liberal arts colleges, new professors face additional challenges. Undergraduates need careful advice on courses and careers – calling on experience that a new faculty member does not have yet. Teaching loads are heavy in comparison with research universities, and physics faculty, in particular, are required to teach many new courses, sometimes ones that are outside their area of direct expertise. Faculty must tailor their research projects to match the needs and abilities of the undergraduates they seek to engage, and the smaller scale of research that small colleges can support.

Experienced mentors can help at this stage by identifying a variety of choices and their possible outcomes, illuminating previously unknown possibilities and helping new faculty begin to carve out their own career path and work out career/ family balance issues. All young faculty members need advice in navigating the tenure process. But each liberal arts college has its own unique "personality". Helping a young faculty member perceive, and thus meet, the particular criteria of excellence at that particular college is something that only mentors familiar with the both the department and the college can do.

Middle Stages

When faculty enter the middle stages of their academic careers, they still need advice and support. What was earlier called "mentoring" is now referred to as "collegial advice" or "peer networks." Faculty now turn to professional peers for advice and support when they write new grants or publish the results of their work. This peer mentoring is very helpful when faculty face changes in their research, teach new courses, or add new pedagogical techniques or technologies to their existing courses. While typical departments will protect pre-tenure faculty from heavy teaching loads or committee duties, at mid-career, large workloads can suddenly surface. Thus, mid-career faculty must make decisions about which committees or projects to support (either in their home institutions or in their professional communities). The need to juggle personal and professional goals continues as family needs change but do not disappear.

A good mentoring network can help faculty make these difficult professional choices. Mentors remind the faculty member how to say 'no' when appropriate, establish priorities, remain flexible, and maintain a balance that respects personal needs and goals, as well as professional ones.

Later Career Issues

Senior faculty face some compelling new issues. Leadership expectations increase while teaching and research demands continue. Research can become more difficult as the momentum that was acquired in graduate school and start-up initiatives have both run out. At a liberal arts college, heavy teaching loads and isolation from research colleagues further complicates the situation. Senior faculty might be chairing a department, or a division of their college – so must balance teaching and research with these important duties. They might also be called upon to serve at the national level.

The needs of adult children and aging parents replace those of infants and young children. The problem of finding gainful employment with a spouse might turn into the problem of separation, divorce, or the loss of a spouse. Senior faculty nearing the end of their regular teaching and research careers face uncertainties in the transition to retirement.

Advice and support from peers is indispensable at later career stages. Peer mentors can encourage each other to try new research avenues and expand their intellectual horizons. They can provide fresh perspective and insight. That is, experienced senior faculty are expected to be the font of wisdom for younger colleagues, but they may be too close to their own situation to apply the same creativity to their own careers.

Horizontal Mentoring

In the broadest and most useful sense, then, mentoring connotes an activity that is not necessarily "top down" or "from one generation to the next." As we've mentioned above, most studies of mentoring are about the effect of faculty mentors helping students, and tenured faculty mentors helping colleagues up through the time of their tenure decision. However, Lederman et al. (*12*) describe a group of four women who were all in non-tenure track positions and formed a mutual mentoring to help them advance their careers. The group focused on professional issues, particularly research. They met frequently, set goals and schedules, read and critiqued each other's work, and gave each other strategic advice. Three of the four achieved tenure; the fourth was offered a job at another institution. They attribute their success to several factors: The small size of the group which created intimacy and made scheduling easier; its single-sex nature which made it more honest and less "academic", the similarity of disciplines and employment histories, and an ethos involving mutual respect.

The premise of these successful NSF Advance alliances is consistent with this study that "horizontal mentoring" between individuals of similar rank, field, academic environment, and perhaps also matched by race and gender, is a highly beneficial enterprise. The roles of mentor and mentee are fluid, and the interaction is beneficial to both.

Whether the people engaged in horizontal mentoring are at an early, middle, or late career stage, the activity of mentoring promotes flexibility and supports the career growth of not only the mentee, but also the mentor. It provides the mentee with external feedback and novel views from another's perspective while challenging the mentor to be flexible in gathering ideas and insights that can be of use to someone in a similar, but not identical, career. This exercise can help mentors identify new options in their own careers.

Mentors both within and outside of a faculty member's home institution have a role. External mentors are extremely helpful at providing a fresh perspective. Internal mentors have specific insights born from knowing the ins and outs of the institution. Also such mentoring relationships create a strong sense of community within an institution.

It goes without saying that having numerous mentoring relationships constitutes a web of support. A horizontal mentoring relationship, even one between just a couple of people, is important. It not only is useful in its own right, but it acts as a basic building block for a larger network - creating additional "nodes" for expansion. Networks provide the varied and sustainable system of support that a single mentor-mentee relationship cannot duplicate. So it is clear that we must build and maintain professional networks, and help our colleagues do the same.

Our Demographic: Senior Women Physics Faculty at Liberal Arts Colleges

The Numbers

Senior women in physics, especially those at liberal arts colleges, face special challenges in trying to develop a network of support. We benefit greatly from mentoring: both top-down and horizontal mentoring, from people who have had similar experiences. But this is often difficult to come by. We generally do not have senior women peers at our own institution who have the relevant experience to guide us. In 2006, 13% of physics faculty at all ranks were women, while 6% of full professors were female (*13*). While there is much that we can learn from our junior colleagues, the issues we face and the demands on our time and energy are often different, as outlined above. Furthermore, society's expectations for women in professional situations are often quite different from the expectations of men (*14*). Our male colleagues, as well intentioned as they might be, are often not able to provide the guidance that we need, when they reason from their own lives and professional experiences. Senior women in other departments, even other science departments, may not fully understand the specific issues that we face as women in physics. We may have been the only woman in the department for years, feeling that we must constantly prove ourselves worthy while "swimming against the tide" of expectations that women are not natural leaders, or equals in the domain of physics. (Four out of five of us were the only woman in the department when they were initially hired; the one of us who was hired into a department with another woman was delighted to know she was hired on her own merits, not because the department needed a woman.) Neither male colleagues nor colleagues in other departments will understand, firsthand, what this entails. It takes time and effort to seek out possible mentors. Relying on serendipity to encounter such people is not practical given our small numbers.

Small College Life

The challenges faced by women physicists at small colleges are often quite different from those faced by women in other academic institutions. We are doubly isolated by being a woman and by being in a department so small that no one works in a related research field. In small departments and small colleges there are often well defined, but not necessarily well-articulated, cultural norms within which we must function. We often need guidance in discerning and adapting to our culture. We may also need guidance in forming strategies for instituting cultural change. Students at private liberal arts colleges may have specific expectations that aren't always consistent with our own expectations as educators, or our previous experiences, when we were students ourselves. It has been shown in numerous studies that women in authority, including academic authority, face expectations that are different from those of men (*15–17*). Students may expect us to nurture and support them in a stereotypically "maternal" way, and will penalize us if we do not. For example, the President of a small liberal arts college tells a story of her first days in office, when a student strode into her office without an appointment, and requested that the new President help her find her lost backpack. She was shocked when the president would not comply. We may also be differentially penalized when we hold students to high standards.

Senior faculty at liberal arts colleges are expected to take on many simultaneous roles in a way that would not be typical at a research university. These demands come not only from within our department or from our research programs. They also come from the college as a whole. Small colleges survive only because some faculty members are willing to put enormous amounts of time into institutional governance, long-range planning, and other projects for which there is little direct reward. We are given little advice about carefully choosing the causes to which we are willing to contribute and saying 'no' to others. As women we may not only be pressed into service more often due to our 'token' status, but we may fall more readily into the trap that if something needs to be done, it is our responsibility to see that it gets done. As senior faculty members, we often feel as if we have to/should be able to "do it all". The role of a mentor with similar experiences is thus to help us set reasonable boundaries, and give us permission to prioritize service projects. This includes giving a priority to time that we take for our personal wellbeing.

Physics versus Chemistry

Senior women in physics and chemistry share many concerns. Our fields are lumped together as "physical science" by many reporting agencies, and for good reason. We share important modes of scientific praxis: a reliance on mathematics, laboratory experiments, and a "theory-rich" knowledge base. For example, one of us (A.B.) , though trained as a physicist, did a postdoc and virtually all of her off-campus research in chemistry departments. From a gender equity point of view, when one looks beyond the bachelor's degree level, there is no environment (grad school, academia, industry, ...) in which one can find equal numbers of women and men in either physics or chemistry. The metaphor of a "leaky pipeline" has been utilized for many decades to describe the outflux of women from both chemistry and physics at critical junctures. One could use this metaphor with confidence as recently as a few years ago, and speak truthfully of women being hired or tenured in both fields at lower rates than men (*18*). Now, recent data indicate that, as a group, women are now more successfully passing careeer milestones, so the pipeline is less leaky. An important qualifier: this refers to non-minority women. For underrepresented minorities, both men and women, the equity situation remains dire (*19*). In summary, a lack of gender balance remains between women and men in both fields at all career levels.

Despite our similarities, there are some differences as well between women physicists and chemists. First of all, there are fewer women in physics percentage-wise. Figure 1 (also see Reference (*20*)) shows that the percentage of women bachelor's recipients in chemistry is roughly 50% nowadays; while it is roughly 25% in physics. If one reads past percentages to the raw numbers, there are significantly fewer women in physics, both as students and faculty members. Women physics faculty are likely to be isolated, especially in a small department, where having a little over 10% women means having one woman. This is not at all an uncommon situation for physics departments; in 2006, approximately 39% of all physics departments had zero or one woman in professorial ranks (*13*). In most small colleges, chemistry departments are larger, so this situation is less frequent.

The lower percentage of women in physics faculty at all ranks at R1 institutions is shown in Figure 2. Interestingly, a smaller proportion of female Ph.D. recipients go into academic positions in chemistry than do in physics (compare Figures 1 and 2, see also Reference (*8*)). The reasons for this are not yet understood (a gender-related decision that favors industry over academia?).

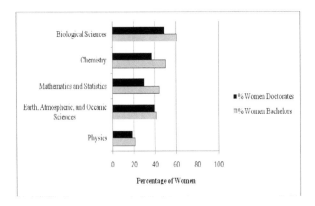

Figure 1. Women degree recipients in the sciences. Data are from Tables C-5, F-1 of the 2007 NSF report, Reference (21).

In terms of our teaching responsibilities and the departmental "climate", there are differences as well. For one, chemistry departments tend to split up the undergraduate curriculum, so that organic chemists do not teach physical chemistry and so on. In physics, by contrast, the model is more often "everyone teaches everything." In this case a young (or not so young) physics faculty member is faced with teaching new courses, most of which are outside her specialty.

Mutual Mentoring and Our Alliance

In this section we describe our alliance—what we did, why we think it has worked so well, and what we consider to be essential. These ideas are summarized in the following list:

- We are similar in age and rank, so our professional and personal issues are similar.
- We are from different institutions and subfields, so we are not in competition.
- NSF funding helped us (and our colleagues) take this project seriously, so it didn't get lost in the myriad other demands on our time.
- The initial face-to-face meeting was essential for us to get to know each other and build trust. Eating meals together created an informal atmosphere.
- Regular Skype calls help us keep in touch with each other's professional and personal lives at no cost.
- Reading *Every Other Thursday* helped us with our initial agenda and gave us some language to identify and discuss our dilemmas and challenges.
- Regular (once or twice a year) meetings maintain our relationships (and are a lot of fun).
- Phone calls and meetings involve discussion of immediate issues and longer-term projects.

The Nuts and Bolts

Given the relatively small numbers of women in physics, particularly at the full professor level, it was no small feat to pull together our horizontal mentoring alliance of women full professors at liberal arts colleges. Establishing a mutual mentoring network across geographic regions and time-zones was non-trivial. What made our group a success in terms of mutual mentoring was a combination of a number of factors, beginning with the alliance structure designed in the NSF-ADVANCE grant. After being (mysteriously) selected and then agreeing to participate in a physics alliance, we began to organize an initial face-to-face meeting. Meeting in person was essential to the success of the group because it allowed us to build a sense of trust and community. This, of course, meant that the grant money used to support travel, housing, and meals was crucial.

We met for the first time at the APS meeting in April of 2008. Some of us had known each other personally or professionally beforehand, but all five of us had not met together before. We were somewhat confused about how to organize ourselves, and structural details (e.g. why we had to eat together to be reimbursed). We began over lunch, by introducing ourselves. In addition to the usual professional information, we included our family situation (We found that all of us had children, though their ages varied from elementary schoolers to 20-somethings.), and a brief discussion of career issues about which we had concerns.

During this conversation and later ones at this meeting, a professional theme emerged; each of us felt uncomfortable about our research. We all realized that we had drifted from our original training into administration, curriculum development, and diversity issues. We were concerned about being "unproductive" compared to physicists at research universities, or to the one or two "superstars" in our home departments. We were wondering if we should try harder to return to doing the kind of work we had been trained to do in graduate school.

All of us, in different ways, were confronting the same series of questions. We agreed to think more about these issues and make that the focus of our next meeting, the following summer. To keep in touch between meetings, we agreed to carve out time for internet conference calls (via Skype) every other week and to begin our alliance essentially as a book group to discuss Ellen Daniell's *Every Other Thursday* (23).

Every Other Thursday, a book about a mutual mentoring network of women scientists, was a good touchstone for our conference calls. In reality, we rarely talked much about the book itself, but it provided a good context in which to share challenges, insecurities, and difficult career issues. It also provided a model of a mutual mentoring network that was useful on our conference calls from logistics—everyone gets a set amount of time to bring up issues and a moderator moves the group along—to language—we started gleefully identifying our own *pigs* (defined by Daniell as "negative self-perceptions").

At our summer 2008 meeting we returned to the theme of our research. We had each worked to define a possible path or paths toward a more satisfactory research career, and to identify some concrete steps we could take along that path. We each took time to describe our possible paths, while others in the group listened

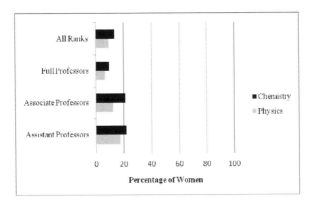

Figure 2. Women faculty in Top 50 research universities. Data are from the 2007 APS Committee on the Status of Women 's Physics Gender Equity Report, Reference (22).

carefully and offered comments and advice. Each person ended with a decision and a series of steps to try. One member of our group expressed the following sentiment:

> I had been working on diversity in physics for about ten years, getting farther and farther from my original field. I love the diversity work, and feel that it's an important contribution to the physics community. But more and more I missed "real physics." I wondered whether I spend my upcoming sabbatical on new diversity projects, or begin a new technical field. The group advised me, if I really missed technical physics, to go ahead and move into a new field. I just finished a very interesting and productive sabbatical, beginning a new research project in atmospheric physics.

Another member of the group decided that her curriculum development work *was* her research—she could continue to develop that work without feeling guilty that she was no longer doing her graduate school experimental work. Another decided to prioritize her own research more.

We met in person again in March of 2009, continuing with conference calls in between. Over that time, the conference calls evolved from book group discussion to helping with the immediate crises of the week as well as some longer term accountability. We shared goals as well as accomplishments. Even though the conference calls were incredibly useful, it was difficult to keep them going and for a while after the March 2009 meeting, we did not conference call regularly again until the fall of 2009 (when we felt under pressure to have a "product" as a result of our alliance).

Why Does It Work?

We had money to travel to meet together, were "forced" to eat together and felt obligated (by agreeing to participate) to gather on conference calls, but this only begins to touch the surface of why the program had such good results for our group. Almost from the beginning, we were willing to open up with each other. Why were we willing to do that? One of the reasons was that we had nothing to lose and everything to gain: none of us would be evaluating the other on an individual campus; none of us would be evaluating or competing with anyone else for grant applications because our sub-fields were too different. While this is likely to have contributed to our success as mutual mentors, we also think that our successful group is something akin to a "resonant phenomenon". The other members of the group "get it" (whatever the issue) right away, without ancillary explanations or elaborate justifications, because they've had similar experiences. They are not just willing to sympathize, but they are truly able to empathize. For many of us, it was our first time in a group with others who are so much like us - or as one member put it: "For the first time I know that it is really not "just me" in the various career issues I have faced." This group was, then, like a sigh of relief. It brought us out of the isolation of being a sole senior woman physicist on campus, and into the warmth and understanding of a group of savvy, senior, women physicists.

Part of the resonant phenomenon was that we were not only allowed, but encouraged to bring everything to the table. We did not have to separate our professional selves from the rest of our life. Our goals and accomplishments extended beyond the purely professional realm. They included things like making a doctor's appointment for a check-up, getting on-line checking established for paying bills, and going to a yoga class regularly. This was a group that not only recognized, but actually required that we include all of our life in the context of career issues. This may have been because our group members had to deal with a number of family crises: children returning home after college, taking care of elderly parents, and becoming a single mother, as well as some dramatic health issues. But even beyond that, we felt encouraged to approach our mutual mentoring more holistically. As one member put it:

> I am allowed to bring all the juggling balls to the table – family-work issues, research-teaching conflicts, community service-personal need balance. I no longer need to juggle the invisible big ball of family needs. I can discuss all my goals and commitments with others who won't scoff so I can bring it all into better balance.

and from another:

> Our alliance has enabled good, healthy career and life choices to "come out of the closet" and become part of my professional life so they can be given the time that they need to be processed.

Out of a resonant phenomenon, our alliance has grown into a supportive network. We have done a lot to support each other in our research work. Our discussions have helped us get clarity in this work as we addressed the questions of what type of research we wanted to be engaged in. We supported individual answers that included switching subfields, staying in the same field, or working in non-traditional research areas (gender issues, curriculum development). Our network helped one of our members in her decision to get back into research within a new field. Through another member she has made a contact for a sabbatical to begin research in this field. We encourage each other to define goals and priorities to help us focus on important projects and be less distracted by lesser demands. We help each other balance our professional and family demands.

We have provided different perspectives when dealing with a particularly troubling issue. For example, one of us mentored a junior woman science faculty who received a negative tenure recommendation from the promotion and tenure committee. As a group, we worked together to help this member decide how to best provide support for her junior colleague and we celebrated with our alliance member when her junior colleague ultimately did receive tenure.

Our alliance has been more helpful to each of us than we expected when we first agreed to participate. As one member put it: "Our conference calls and meetings are intellectually invigorating as well as one of my most valued sources of wisdom, support and encouragement." While another member says: "Our alliance has given me the courage to accept challenging leadership and difficult tasks because I know I have a backup group for brainstorming and support."

Future?

Now that we know how professionally and personally beneficial our mutual mentoring has been, will we still be able to sustain our already established mentoring network once the NSF support expires? Does a group like ours need a funding source to continue? And most importantly, what are the implications for future mentoring networks?

Perhaps creation and support of such mutual mentoring networks would not require enormous amounts of external funding. If NSF or professional society support could be used to establish methods for creating mutual mentoring networks, then professional societies might be able to provide space at national meetings to allow these mentoring groups to have face-to-face meetings. In physics, for example, the Committee on the Status of Women in Physics (CSWP) currently has funds to run and provide support for professional development workshops for women held in advance of APS national meetings. The workshops have been structured to address the needs women physicists at particular stages of their careers. Structured mentoring networks could be a natural outgrowth of these workshops: a place for the necessary initial face-to-face meetings with their new mentoring cohort.

With modest financial support, members of these small cohorts could gather on a yearly basis when they attend future society meetings. A whole host of social networking technologies and methods of electronic communication could be used to provide more regular "conferencing at a distance". Women who once faced isolation and lack of collegial support and advice would then have the opportunity to participate in a mentoring cohort. Members of these small cohorts could meet periodically in a larger forum to share and collect ideas they have generated , and allow for the professional society to develop action items to address common themes. The IUPAP international conferences for women in physics, three of which have been held so far (*24*), are an admirable model for developing action items. Their conference proceedings show the emergence of common themes, and contain recommendations for institutional and governmental change.

With NSF and professional societies promoting the establishment of these networks, mutual mentoring could be given the professional recognition it deserves. The credibility of an NSF project, or formal recognition within the auspices of the field's professional society, is important. It is much easier to justify mutual mentoring conference calls if it is for an "NSF project" than for a "Women in Physics" group. Currently NSF grants require proposers to specify means of assessment and methods of dissemination of project results. Why not also acknowledge the need for effective mentorship to promote creativity and inspire the future generation of scientific questions and research?

As Skype announced that the alliance call was beginning, I thought about the pile of labs I needed to grade, my unfinished grant proposal, and the committee minutes I still needed to write up and thought 'do I really have time for this conference call?' I answered anyway, heard someone sigh and say she was swamped, and I was reminded once again that I wasn't alone... This was definitely worth it.

References

1. Karukstis, K.; Gourley, B.; Wright, L.; Rossi, M. *Horizontal Mentoring Alliances to Enhance the Academic Careers of Senior Women Scientists at Liberal Arts Institutions*; National Science Foundation PAID-ADVANCE Grant; 2006.
2. Boyle, P.; Boice, B. *Innovative Higher Educ.* **1998**, *22* (3), 157–179.
3. Luna, G.; Cullen, D. *ASHE−ERIC Higher Educ. Rep.* **1995** (3), 1–87.
4. Olmstead, M. *CSWP Gazette* **1993**, *13* (1), 8–11.
5. Ziman, J. *Real Science: What It Is and What It Means*; Cambridge University Press: Cambridge, U.K., 2000; Chapters 2 and 3.
6. Wennerås, C.; Wold, A. *Nature* **1997**, *387*, 341–343.
7. Blau, F.; Currie, J.; Croson, R.; Ginther, D. *Economic Review, Papers and Proceedings*, 2010, in press.
8. *Gender Differences at Critical Transitions in the Careers of Science, Engineering and Mathematics Faculty*; National Research Council: Washington, DC, 2010.
9. Etzkowitz, H.; Kelmelgor, C.; Uzzi, B. *Athena Unbound: The Advancement of Women in Science and Technology*; Cambridge University Press: Cambridge, U.K., 2000.
10. Whitten, B. *Different Scientists, Different Science?* 2010, in preparation.
11. Nolan, S.; Buckner, J.; Marzabadi, C.; Kuck, V. *Sex Roles* **2008**, *58*, 235–250.
12. Lederman, M.; LaBerge, A.; Zallen, D. *Gates* **1994**, *1*, 26–31.
13. Ivie, R. *Women in Physics and Astronomy Faculty Positions*; Statistical Report; American Institute of Physics: Melville, NY, 2006. http://www.aip.org/statistics/trends/highlite/women3/faculty.htm (accessed March 30, 2010).

14. Valian, V. *Why So Slow? The Advancement of Women*; MIT Press: Cambridge, MA, 1998; Chapter 7.
15. Sinclair, L.; Kunda, Z. *Pers. Soc. Psychol. Bull.* **2000**, *26*, 1329–1342.
16. Rudman, L. A.; Kilianski, S. E. *Pers. Soc. Psychol. Bull.* **2000**, *26*, 1315–1328.
17. Hall, L. E. *Who's Afraid of Marie Curie?*; Seal Press: Emeryville, CA, 2007; Chapter 4.
18. National Academies (National Academy of Science, National Academy of Engineering and Insitute of Medicine). *Beyond Bias and Barriers*; National Academies Press: Washington, DC, 2006.
19. Ivie, R. *Lessons Learned from Data on Women's Careers in Academic Physics*; American Physical Society March Meeting, Portland, OR, 2010; Data from *Gender Differences at Critical Transitions in the Careers of Science, Engineering, and Mathematics Faculty*; National Research Council Report; 2009.
20. Leggon, C. B. *J. Technol. Transfer* **2006**, *31*, 325–333.
21. *Women, Minorities, and Persons with Disabilities in Science and Engineering*; Statistical Report; National Science Foundation: Arlington, VA, 2007. http://www.nsf.gov/statistics/wmpd/pdf/ nsf07315.pdf (accessed June 15, 2010).
22. Committee on the Status of Women in Physics (CSWP). *Gender Equity: Strengthening the Physics Enterprise in Universities and National Laboratories*; American Physical Society: College Park, MD, 2007. http://aps.org/programs/women/workshops/gender-equity/upload/ genderequity.pdf (accessed March 30, 2010).
23. Daniell, E. *Every Other Thursday: Stories and Strategies from Successful Women Scientists*; Yale University Press: New Haven, CT, 2006.
24. *Women in Physics: Third IUPAP International Conference on Women in Physics*; Hartline, B.; Horton, R.; Kaicher, C. M., Eds.; American Institute of Physics: Melville, NY, 2009.

Chapter 9

Women Chemists Web: Building Strength through Connections

Carol Ann Miderski*

Department of Chemistry, Catawba College, Salisbury, NC 28144
***cmidersk@catawba.edu**

Chemistry Departments at four-year colleges vary widely in the number of faculty and their expectations regarding the balance of teaching and research. Out of 30 schools within approximately 100 miles of Catawba College in central North Carolina, chemistry department sizes vary from one to nine faculty members with an average of three. Of these colleges, 23% had no women faculty and 53% had only one. Under these circumstances, women faculty often find themselves in an isolated position where they are the only one teaching in their discipline and also the only woman in the department. The Women Chemists Web was initiated in 2009 to bring women faculty from regional colleges together to get to know each other and to develop a resource network. The group is designed to serve as a source of outside perspective, fresh ideas, and alternative strategies for facing the academic, professional and personal challenges encountered in small college environments.

Introduction

Small colleges provide a rich and nurturing environment for students seeking small class sizes and one-on-one interactions with faculty. Many students find this more protected environment less intimidating than larger university settings and more suited to their temperaments and current developmental stages. Faculty who are drawn to teach chemistry within these institutions are also seeking a different balance of professional responsibilities with a distinct emphasis on teaching. The individualized attention that is the hallmark of small colleges comes with distinct professional burdens. Teaching in this environment is an exercise in adaptation

© 2010 American Chemical Society

and problem-solving. Between preparing and running labs, teaching classes, and grading papers, faculty also need to serve on committees, fix equipment, meet visiting families, and direct student research. As a faculty member in a small college, you are called upon to be a part of the fuller educational experience of the campus including cross-disciplinary teaching, academic advising, supervision of student organizations, and community outreach. In spite of the challenges, teaching in the small college environment can be incredibly rewarding as you help to nurture your students throughout their academic careers.

As women chemists in small departments, we face isolation on many fronts. Most schools have only one chemist from each sub-discipline and few have more than one or two women. The value of professional networks has long been demonstrated by organizations such as the American Chemical Society (ACS) (1), the Women Chemists Committee (WCC) (2), the American Association of University Women (AAUW) (3), and the Association for Women in Science (AWIS) (4). Research indicates that mentoring can have many positive effects including improved self-confidence and job satisfaction while fostering professional and personal development, with informal mentoring being more effective than formal mentoring arrangements (5). Peer relationships also provide many mentoring functions with the added advantage of all parties receiving benefits of improved sense of competence and development of professional identity (6). While men and women all benefit from mentoring and networking opportunities, conversation among women tends to be more personal in nature (7) yielding greater benefit to them in relation to issues such as family, health, and interpersonal dilemmas. The need to maintain professional credibility may inhibit discussions of personal issues such as home and work balance in conversations where men are present.

Existing regional organizations for chemists such as local sections of the ACS and WCC are valuable resources but do not directly address the diverse needs of women faculty. This paper will discuss efforts to develop a resource network among women chemistry faculty at four-year colleges in central North Carolina. The primary purposes of the network are to reduce professional isolation and to develop relationships among participants for mutual support, mentoring, and sharing of resources and information. To evaluate opportunities for constructive collaboration within the network, this paper will also explore the professional profiles and concerns of the women participants and the demographics of the institutions and departments they represent. Additionally, interests of participants in various types of network events or activities will be discussed.

Why Build Connections?

Given that time is often the most precious commodity for women chemistry faculty at small colleges, why should we spend it on building connections with faculty from other institutions? After all, we certainly have enough to do with the responsibilities of our own jobs and personal lives. The benefits of developing these connections need to be substantial in order to justify spending our time and efforts. The benefits obtained from connecting with faculty at different institutions are perhaps most undervalued in the small college environment. It is very easy in this environment to get swallowed up in day-to-day responsibilities and assume that your situation is unique.

For faculty at small colleges, the perennial focus is on improving student learning. Developing connections with other chemistry faculty in similar institutions can have a direct impact on student learning in our home institutions. In conversation with one another, we can share effective teaching strategies and improve our workload management by increasing our awareness of successful models implemented by others in similar work environments. Issues such as classroom management and the burdens of grading and assessment are universal throughout academe. Development of ties with peers can reduce professional isolation, inspire us to take on new challenges, and empower us to say no to excessive or inappropriate demands on our time. Additionally, these professional connections can provide access to resources and expertise such as grant, publication, conference or collaboration opportunities and as well as inspirations for course topics or lab development. Interactions with other chemistry faculty can also increase research productivity by providing a sounding board for ideas and successful models for making time to do more research.

When faculty have adequate external peer support, their institutions benefit as well. As faculty rotate through department chair positions, connections to their peers provide a rich source of experience to draw upon to solve personnel and managerial problems. Often, an outside perspective is critical in finding a way to recognize and break out of non-functional patterns that have evolved over time. Collectively, a peer resource network has access to faculty who have served in most academic governance roles. As institutions face the need to develop policies or engage in curricular reform, knowledge of effective solutions elsewhere can often provide an excellent starting point for creating a local solution. While each institution is unique, the basic framework of academic institutions is similar enough to provide useful models. When programs or institutions are undergoing external evaluation or accreditation, a peer network can provide a source of outside reviewers, useful feedback on assessment strategies, or ways to present the value of departmental efforts more effectively.

Most professionals face challenges when trying to maintain a balance between the demands of home and work. For women chemists in small departments, finding the balance is even more complex. In the university environment, another person who has taught the same course can often be found just a few steps down the hall. In a small college environment, you may not know another person who could teach your courses in your geographical area. This isolation provides special challenges when illness, family care, or sabbaticals necessitate finding a short term replacement. If women science faculty are rare in a particular institution, there may not be effective mechanisms to handle the unique challenges of maternity or child care issues for faculty who teach laboratory courses. Access to women chemistry colleagues provides opportunities to see how others have negotiated the difficulties of finding balance between professional and family responsibilities. Gender also plays a role in classroom dynamics and campus politics. It can be a tremendous asset to have a peer who can serve as a sounding board from outside the institution who has negotiated similar challenges in the same type of working environment.

Methods

The Women Chemists Web was initiated as a component of my participation in the National Science Foundation ADVANCE-PAID project, "Collaborative Research for Horizontal Mentoring Alliances." (8) Participants in that project were organized into five member Alliances designed to study and develop horizontal mentoring systems for senior women chemistry faculty. Within the Alliance structure, conversations with other women in similar professional circumstances were of tremendous personal and professional benefit. One of the tasks of our group was to initiate an event in our local environments. Since I received the greatest benefit from meeting others and having a relatively unstructured opportunity to talk about the challenges and successes of working in a liberal arts environment, I thought it would be beneficial to reach out to other women faculty in my area.

Women Chemists Web

The Women Chemists Web was developed to provide a resource network for women chemistry faculty providing opportunities for information sharing and development of personal connections among faculty in different institutions in the central region of North Carolina. The name for the organization evolved from a composite graphic, shown in Figure 1, designed for an invitation to the first meeting.

The concept of network development led to the web image which also resonated with the interconnectedness of our many responsibilities as women faculty and the impossibility of separating the personal from the academic and professional. The dangling raindrops reinforced the concept of potential fallout when we make adjustments in our lives.

The first step in developing the network was to identify women in the area. Utilizing the College Board College MatchMaker function (9), four-year private institutions within 100 miles of Catawba College in Salisbury, North Carolina were identified. The list included some schools without chemistry departments, such as divinity and business schools. A small number of additional schools were included that were not identified by the College Board search. The websites of all schools were used to identify how many faculty were in the chemistry department and if any were women. In schools that had Natural Science departments, faculty with chemistry specialties or chemistry teaching responsibilities were included.

An e-mail was sent to all women chemistry faculty inviting them to an initial meeting of the Women Chemists Web at Catawba College. The initial meeting was designed to provide an opportunity for women to get acquainted and discuss issues of interest. The women were also invited by e-mail to participate in an on-line survey using Survey Monkey (10) to gather resource information to be shared with other participants. The first section of the survey requested basic demographic information to be shared with others in the Resource List including personal information such as teaching specialties, rank, and years of service and departmental information such as teaching load, research expectations, and number of graduates. The second section surveyed the importance of various

Figure 1. Women Chemists Web graphic.

academic, professional, and personal topics. This was followed by a section on the types of events of interest and an opportunity to provide general comments. The demographic and contact information was compiled into a Resource List that was shared with all women who completed the survey. The Resource List enables participants to identify specific women for direct contact or to request input from the full group.

Results and Discussion

Institutional Profiles

Based on the search for four-year colleges, thirty private four-year colleges were identified with chemistry departments varying in size from one faculty member to nine. Student enrollments varied from 750 to 5000. Included within the group were several women's colleges and historically black colleges. Academic profiles varied widely from open admission to highly selective with some campuses being residential and others having substantial commuter populations. Some schools were research intensive while others had only limited research expectations.

The majority (77%) of chemistry departments had three or fewer faculty members. Three faculty was also the mean, median and mode of department size. Based on information gathered from department websites, women comprised 34% of all chemistry faculty in these institutions. The representation of women at individual schools is summarized in Table I.

Table I. Representation of Women in Chemistry (30 schools)

Department Composition	Percent of Departments
NO women chemists	23%
ONE woman chemist	53%
TWO women chemists	20%
THREE or more women chemists	1% *women's college

The number of chemistry departments with small programs was greater than expected, highlighting the need for professional networking for those teaching out of field. With more than half of the departments having only one woman chemist, the isolation is compounded. An unfortunate side effect of this demographic is the number of future scientists who will have encountered only one female role model in the physical sciences, if any.

While not directly tied to the focus of this paper, the prevalence of very small chemistry programs as illustrated in this sample raises concern for the future of chemical education in small colleges. These institutions are strongly impacted by the economic realities of the expense per chemistry student and growing trends towards reducing the general education requirements in science. Revisions in MCAT requirements (*11*) will necessitate some restructuring of chemistry programs. Chemistry departments at small colleges are at serious risk of being reduced to service programs for other departments, much like the earlier fate of many small college physics programs. It is essential that the chemistry community work together to promote survival of smaller programs. These programs are often highly successful at developing students who find larger institutions too intimidating when they are fresh out of high school. We cannot afford to have the only national choice for chemistry majors restricted to large university settings.

Department Profiles

The chemistry departments represented ranged broadly in size and service load. Advanced courses are offered annually in half of the schools, presumably on an alternating year basis in the others. Access to support staff varied widely with some departments using work-study students to help with laboratory setup and others having a professional lab manager who was often shared with other departments. Many respondents mentioned the presence of a shared administrative assistant. While nearly all departments required or encouraged research both for faculty and students, only half required or encouraged sabbaticals. For many, access to travel funds was limited or available only on a competitive basis. The faculty contact hours reported through the survey ranged from nine hours (7%) to fifteen hours (29%) and the majority reported twelve hours (64%). According to the ACS-Committee on Professional Training Guidelines (*12*), fifteen hours is a maximum teaching load and should not represent normal teaching loads.

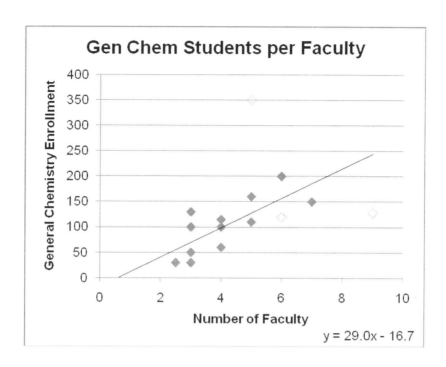

Figure 2. General Chemistry enrollment compared to department size.

On most campuses, the largest teaching load is associated with the delivery of General Chemistry. The relationship between the enrollment in General Chemistry and the number of chemistry faculty is shown in Figure 2.

In spite of General Chemistry enrollments varying by a factor of ten or more in colleges with widely differing entrance requirements, a fairly linear ($R^2 = 0.6$) relationship among departments is observed. As might be expected, the service load of the department measured through the enrollment in General Chemistry appears to have a noticeable effect on the number of faculty needed. The relationship implies that a sustained increase of 30 students in the enrollment for General Chemistry may merit increasing the departmental size by one faculty position. The three institutions indicated by open marks in Figures 234 were not included in the linear fits. These departments were found to be outliers in at least two of the three relationships studied. The departments in question have distinctive institutional characteristics which might reasonably cause their differences from the rest of the group. Discussion of these distinctions is not included as it would reduce the anonymity of the sample.

Increasing enrollments in General Chemistry may reasonably be assumed to lead to higher enrollments in subsequent courses and an increase in graduating chemistry majors. Figure 3 shows the relationship between the number of chemistry majors and enrollment in General Chemistry.

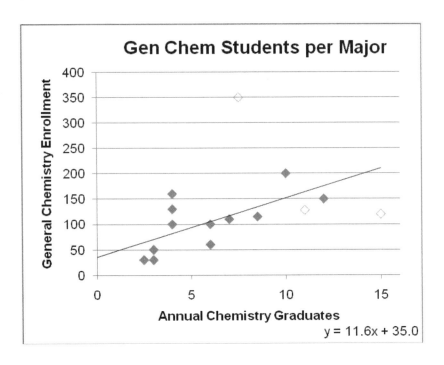

Figure 3. General Chemistry enrollment compared to chemistry graduates.

Though the institutional types vary widely, there is also a linear relationship between General Chemistry enrollment and the number of chemistry majors graduating annually, though the relationship is quite weak ($R^2 = 0.4$). Most schools experience wide shifts in the number of chemistry graduates each year. When the number of graduates was reported as a range, the midpoint was used. In very general terms, there appears to be a threshold size General Chemistry enrollment needed to be able to cultivate chemistry majors. Increasing the size of the entering General Chemistry class also increases the number of graduates that can be anticipated.

The delivery of advanced courses and opportunities to pursue research projects places a substantial burden on faculty. The relationship between the annual number of chemistry graduates and the size of the Chemistry Department is shown in Figure 4.

As might be expected, the department size is also impacted by increasing numbers of chemistry majors. This relationship is the strongest ($R^2 = 0.8$) which might be expected due to the percentage of faculty teaching load dedicated to teaching upper level courses. Beyond a threshold, the slope implies that a sustained increase of two chemistry graduates per year may necessitate the addition of a faculty position. Given how widely the number of majors may vary on a year-to-year basis, convincing the administration to add a position can be expected to be challenging.

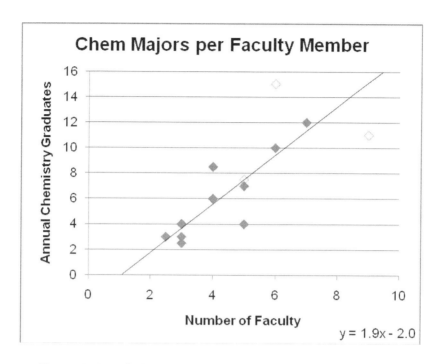

Figure 4. Annual chemistry graduates compared to department size.

Faculty Profiles and Concerns

Approximately half of all women invited to join the Women Chemists Web elected to respond to the survey and all came from departments of at least three faculty members. Those in the smallest departments may be more effectively reached during the summer months when schedules allow more breathing space. Additional efforts are underway to reach out to those who chose not to respond. The 23 women who did respond represented 18 schools. Their teaching experience ranged from one to thirty years with a mean value of fourteen years. Thirteen percent were in their first six years of teaching, 30% had 7-13 years experience and 57% had 14 or more years of teaching experience. These numbers were somewhat consistent with the distribution of rank, 13% Assistant, 26% Associate, and 61% Full Professors. One area for potential concern is that 29% of women with 14 or more years experience had not yet been promoted to the rank of full professor. The distribution of teaching specialties included 29% biochemistry, 24% organic, 19% physical, and 9% analytical chemistry, with the remainder in inorganic, environmental and general chemistry.

One of the purposes of the survey was to determine the issues most important to women chemistry faculty. Participants were asked to rank the importance at this point in their careers of thirty-one topics. Choices ranged on a four-point scale from very important to not important with an additional option for not relevant. The topics included are categorized and shown in Table II sorted by the percent of all "very important" and "moderately important" responses.

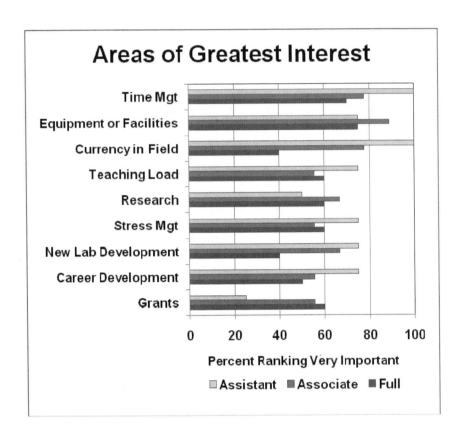

Figure 5. Areas ranked very important by half of respondents, sorted by rank.

As a group, respondents showed greater interest in Academic and Personal topics. Interest tended to be lower for topics that relate to shorter periods within the working career such as tenure, chair responsibilities, child care and maternity.

Nine topics received a "very important" rating from more than half of the respondents and the distribution by faculty rank is shown in Figure 5.

Of the nine areas reported here, most responses do not differ substantially by faculty rank. The two exceptions are currency in field, which becomes less important to faculty with promotion in rank, and concerns with grants, which increase after the assistant rank. Other areas which are not shown had strong responses for one rank: home and work balance (75%) was a concern primarily at assistant rank along with tenure (50%), promotion was of concern for assistants (50%) and associates (44%), while sabbatical (50%), retirement (40%), and chair responsibilities (40%) were most salient for full professors. When both "very important" and "moderately important" responses are grouped together, interest levels rise to greater than 90% for all nine categories in Figure 5 with the exception of research (87%) and grants (74%), primarily due to lower interest among full professors.

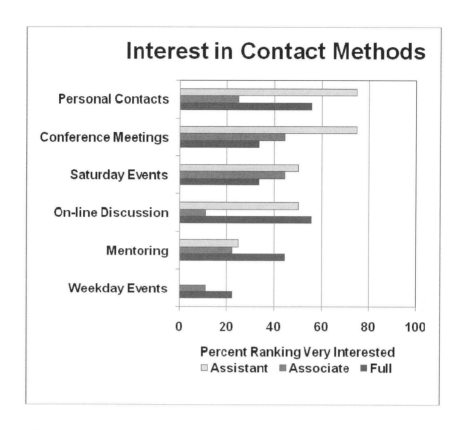

Figure 6. Events or activities ranked very interesting, sorted by faculty rank.

Survey participants were also asked to rank their interest in the types of events or activities that could be offered by the Women Chemists Web. Their responses are summarized in Figure 6.

Of the possible contact methods proposed, weekday events were least favored with none of the assistant professors indicating they were very interested. This feedback is consistent with comments regarding the timing of our first event, which was held on a Friday afternoon. When "very interested" and "moderately interested" responses were grouped together, interesting differences were observed by faculty rank. Assistant and full professors showed the highest overall interest, with personal contacts, on-line discussion, and mentoring categories above 90%. Weekday and Saturday events were next at 75%. Associate professors indicated less than 67% interest for all categories except conference meetings (78%). The lowest interest level for associate professors was in on-line discussion (44%) and for full professors was in conference meetings (56%). Mean values of interest for all types of contact methods were 87.5% for assistants, 60% for associates, and 80% for full professors.

Table II. Survey Topics with Percent Responses "Important"

Academic		Professional		Personal	
Teaching Load	96	Career Development	91	Time Management	96
Equipment/Facilities	95	Research	87	Stress Management	91
Currency in Field	91	Budget	78	Home/Work Balance	91
Lab Development	91	Grants	74	Personal Health	91
Course Development	91	Sabbatical	74	Family Health	91
New Pedagogy Access	68	Administrative Duties	65	Family Activities	61
Enrollment	65	Promotion	57	Marital Status	59
Teaching Out of Field	64	Retirement	55	Spouse Employment	50
Student Behavior	50	Office Politics	50	Child Care	35
		Tenure	39	Elder Care	27
		Chair Duties	35	Maternity	27

Connections

Since its inception in October of 2009, participants in the Women Chemists Web and Resource Network have participated in a variety of activities. The first meeting was a Friday afternoon reception held at Catawba College. While only six faculty attended, the small number was perfect for getting to know one another and having in-depth discussions on topics of mutual interest. Conversations were wide-ranging, covering topics as diverse as retirement planning, travel courses, family issues, and tenure procedures. A similar experience was provided by a dinner gathering with nine members, though the ambient noise level of the restaurant kept conversations to smaller groups. One essential component to reaping the greatest rewards from this type of networking is the opportunity to have loosely structured, face-to-face interactions. Women chemists are inherently problem-solvers and when one person discusses challenges or frustrations in her own situation, others inevitably are ready to provide alternative solutions or approaches from their own experience. Also, participants can find more effective approaches to common problems without first having realized their own approach was in need of re-evaluation. One great advantage provided by networking is raising awareness of when it is important to challenge the status quo while providing successful strategic models. Information about procedures and policies at other institutions can provide necessary leverage to effect change at home by providing examples that can be shared with administrators. This approach is especially effective when examples can be cited from either peer or aspirant group institutions.

The Resource List has also been useful to members for target-specific problem solving. One member used the list to solicit information regarding chemistry library holdings to take back to her home institution. Another member used the list to disseminate a posting for a job opening. The list has provided a mechanism for making members aware of presentations or symposia at national and regional meetings, allowing us to support one another through our attendance. Members have also taken advantage of opportunities to get acquainted at conferences in a one-on-one setting. Contact information from the Resource List was used to provide a professional development opportunity for three women who were invited to participate in the recent NSF sponsored Summit on the Advancement of Senior Women Scientists at Liberal Arts Colleges. E-mail communications among network participants provide rapid feedback that may address specific individual, departmental, or institutional needs.

The professional connections developed through the Women Chemists Web constitute a valuable resource for personal and family issues as well. As the data demonstrate, many women chemists are alone in their departments. A network of other women who truly understand the challenges of women faculty in small chemistry departments can be of incredible value to the woman facing issues of work/family balance. Whether the issues are being single in a small college town, ways to handle pregnancy while in the lab, getting kids to after-school programs, or finding solutions for the care of aging parents, having colleagues who have faced similar challenges in the same type of environment can provide alternative approaches and strategies. It is easy to underestimate the reduction of stress provided simply by the knowledge that you are not alone. When it feels like everyone in your life needs a piece of you, it is wonderful to have a colleague who can serve as your personal flight attendant and remind you to "position your own mask before assisting others."

Future Connections

Feedback from participants in Women Chemists Web activities has been strongly positive and indicates interest for increased involvement. Several faculty are working on initiating a metropolitan area chapter of Iota Sigma Pi, the National Honor Society for Women in Chemistry (13). Not only will this enhance the connections among the women faculty, it will also set a good model for our women students and help them start their own networks. Efforts are also underway to start an on-line community utilizing a social networking site provider such as Ning (14). A site with a community bulletin board, discussion groups and e-mail services could provide a centralized way for Women Chemists Web members to stay informed and reach out to one another. Women who have heard about our group at conferences have already inquired about joining. Given that the group was initiated less than one year ago, the interest level and response to activities and information requests indicate that women chemistry faculty throughout the region are committed to *Building Strength through Connections*.

Acknowledgments

This work would not have been possible without the opportunity to participate in the NSF ADVANCE-PAID project on Collaborative Research for Horizontal Mentoring Alliances [NSF-HRD-061840, 0619027, 0619052, & 0619150]. My thanks go to Bridget Gourley of DePauw University and Kerry Karukstis of Harvey Mudd College for convincing me that participating would not be just more work to do. I also would like to thank my fellow Alliance members: Ruth Beeston of Davidson College, Jill Granger of Sweet Briar College, Darlene Loprete of Rhodes College, and Leslie Lyons of Grinnell College. Working with our Alliance has been pivotal in revitalizing my professional life, redefining my workload, and helping me find more effective balance while learning when and how to say no.

References

1. ACS Network Landing. http://portal.acs.org/portal/acs/corg/networkLanding?_nfpb=true&_pageLabel=PP_MNLANDING&node_id=2127&use_sec=false&__uuid=1fc99b8c-ca73-45c7-b608-b4c8152df856 (accessed June 15, 2010).
2. WCC-Networking. http://womenchemists.sites.acs.org/wccnetworking.htm (accessed June 15, 2010).
3. Connect with AAUW. http://www.aauw.org/connect/ (accessed June 15, 2010).
4. AWIS Community. http://www.awis.org/displaycommon.cfm?an=5 (accessed June 15, 2010).
5. Ragins, B. R.; Cotton, J. L. Mentoring functions and outcomes: A comparison of men and women in formal and informal mentoring relationships. *J. Appl. Psych.* **1999**, *84*, 529–550.
6. Kram, K. E.; Isabella, L. A. Mentoring alternatives: The role of peer relationships in career development. *Acad. Manage. J.* **1985**, *28*, 110–132.
7. DeCapua, A.; Berkowitz, D.; Boxer, D. Women talk revisited: Personal disclosures and alignment development. *Mutilingua* **2006**, *25*, 393–412.
8. *Collaborative Research for Horizontal Mentoring Alliances*; National Science Foundation ADVANCE-PAID Project, NSF-HRD-061840, 0619027, 0619052, 0619150; 2006.
9. College MatchMaker. http://collegesearch.collegeboard.com/search/adv_typeofschool.jsp (accessed June 15, 2010).
10. Survey Monkey. http://www.surveymonkey.com/ (accessed June 15, 2010).
11. *Scientific Foundations for Future Physicians*; Report of the AAMC-HHMI Committee; American Association of Medical Colleges: Washington, DC, 2009; pp 8–13. http://www.hhmi.org/grants/pdf/08-209_AAMC-HHMI_report.pdf (accessed June 15, 2010).
12. *Undergraduate Professional Education in Chemistry*; American Chemical Society: Washington, DC, 2008; p 4. http://portal.acs.org/portal/PublicWebSite/about/governance/committees/training/acsapproved/degreeprogram/WPCP_008491 (accessed June 15, 2010).

13. Iota Sigma Pi–National Honor Society of Women in Chemistry. http://www.iotasigmapi.info/ (accessed June 15, 2010).
14. My Ning Networks. http://www.ning.com/networks (accessed June 15, 2010).

National Initiatives

Chapter 10

Development of a Horizontal Peer Mentoring Network for Senior Women Chemists and Physicists at Liberal Arts Colleges

Kerry Karukstis,[*,1] Bridget Gourley,[2] Miriam Rossi,[3] Laura Wright,[4] and Anne-Barrie Hunter[5]

[1]Department of Chemistry, Harvey Mudd College, 301 Platt Blvd., Claremont, CA 91711
[2]Department of Chemistry, DePauw University, 602 S. College Ave., Greencastle, IN 46135,
[3]Department of Chemistry, Vassar College, 124 Raymond Ave., Box 748, Poughkeepsie, NY 12604
[4]Department of Chemistry, Furman University, 3300 Poinsett Highway, Greenville, SC 29613
[5]Ethnography & Evaluation Research, University of Colorado, Boulder, 580 UCB, Boulder, CO 80309
*Kerry_Karukstis@hmc.edu

Our research project focuses on the distinctive environments of undergraduate liberal arts institutions and the challenges faced by senior women faculty on these campuses to attain leadership roles and professional recognition. The project involves the formation of five-member alliances of senior women faculty members at different institutions for the purpose of "horizontal mentoring." Three of the alliances are comprised of full professors of chemistry, the fourth involves full professors of physics. We have formed these alliances to test a "horizontal mentoring strategy" that aims to enhance the leadership, visibility, and recognition of participating faculty members. Alliance members participate in discussions, workshops, and activities focused on career and leadership development through periodic gatherings of alliance members at various locations across the country and through electronic communication via online collaboration tools. The alliances are

© 2010 American Chemical Society

networked to augment the peer-support structure with a larger cohort of senior women scientists. Outreach activities on home campuses extend the impact of the career development expertise attained by project participants. This NSF-ADVANCE-PAID project is also working to identify and create resources that address career development issues for senior women at liberal arts institutions and disseminate best practices on horizontal mentoring strategies for academic women. We have explored how our institutional structure and culture can profoundly influence the career challenges of academic women and how our mentoring strategy can operate particularly effectively for women from liberal arts colleges.

The Value of Mentoring for Senior Women STEM Faculty Members

The underrepresentation of women in almost all physical science and engineering fields is a well-documented statistic (1). One strategic effort to broaden the participation of women in the technical workforce is to increase the advancement of women faculty in science and engineering fields at academic institutions of higher learning. The presence of senior women faculty in the highest ranks of academic leadership enables female students to self-identify as potential scientists and engineers, thus having a powerful influence on their choice of career. Yet the percentage of women in senior faculty positions in science and engineering is discouragingly low. For example, only 1.5% of those faculty members in universities and four-year colleges holding doctorates in the physical sciences are female full professors with 20 years or more experience beyond their doctorate (2). In contrast, the corresponding figure for male full professors with the same level of experience and doctoral field is 21.8%.

Analysis of the disproportional presence of senior women scientists and engineers at colleges and universities is a complex and multifaceted process. Numerous variables have been examined for their impact on advancement in academia. Compelling evidence exists to support the hypothesis that both formal and informal mentoring practices that provide access to information and resources are effective in promoting career advancement, especially for women (3, 4). Such associations provide opportunities to improve the status, effectiveness, and visibility of a faculty member via introductions to new colleagues, knowledge of information about the organizational system, and awareness of innovative projects and new challenges (5–8). Some of the specific benefits accorded to mentees compared with their colleagues with no mentoring support include enhanced socialization to institutional and professional organizations; greater productivity measured in terms of research, grants, and publications; and increased recognition from colleagues and authorities in the field (9–11). Given these favorable outcomes, higher education institutions, many funded through NSF-ADVANCE Institutional Transformation Awards (12), have established mentoring programs to improve the campus climate for women faculty and facilitate their retention and promotion.

While mentoring is traditionally viewed as essential early in one's career, the changing responsibilities of faculty members as they advance in the professoriate suggest that mentoring relationships would also facilitate career advancement for senior faculty seeking new challenges and leadership roles and desiring greater professional visibility and recognition. Endowed professorships, department chair positions, membership on tenure and advancement committees, or roles as associate deans or chairs of the faculty are common leadership opportunities for senior faculty. In addition, as a consequence of particular expertise acquired throughout a career, senior faculty might also consider other leadership activities both at their institution and at a national level, including directors of interdisciplinary programs, offices in professional organizations, chairs of national disciplinary and professional conferences, or roles as principal investigators on multi-institutional collaborative grants. All of these career ambitions for senior faculty reflect goals of institutional and national leadership in the profession and are challenging aspirations. The coaching and advice of experienced mentors would be valued resources to assist a faculty member in achieving these goals.

Just as in the case of prospective women scientists, senior women faculty often prefer mentors who are like themselves because they perceive such female role models to have experienced professional and personal difficulties and challenges similar to their own (13–15). Yet, as there are few women faculty in high-ranking positions, cross-gender mentoring is likely to be the only "traditional mentoring" option available for senior women faculty in science and engineering. To remedy the lack of access to experienced female mentors, alternative models of mentoring must be found and research conducted to better understand the benefits of these different forms of mentoring in academe.

Critical Needs for Senior Women Chemistry Faculty at Liberal Arts Institutions

The ambition of senior women chemists at liberal arts institutions to attain the highest leadership positions at their institutions as well as in national professional organizations is a challenging goal. Private, residential liberal arts colleges are typically characterized by strong faculty governance, strong expectations of service, an emphasis on teaching with small classes and low student-faculty ratios, and small departments with few colleagues in one's area of specialization. These settings provide both unique career growth opportunities and challenges for women faculty members. Using local resources to identify female career mentors in the discipline is a limited option. Indeed, the dearth of senior women faculty in chemistry is even more acute at B.S.- and B.A.-granting institutions than at Ph.D.-granting universities. The Women Chemists 2000 publication of the American Chemical Society (16) reported only 932 female full professors of chemistry at baccalaureate institutions compared with 1696 at PhD-granting institutions. With 2.3 times as many baccalaureate institutions than Ph.D.-granting institutions in the United States (647 vs. 283, as reported in the most recent version of The Chronicle of Higher Education Almanac (*17*)), the scarcity and consequential isolation of senior women chemists at such undergraduate institutions is profound. Alternative modes of mentoring are a necessity if a strong support system is to be established along gender-specific lines to improve the climate for senior women chemists at private liberal arts institutions and to facilitate their advancement to leadership positions.

Formation of Our Inaugural Horizontal Peer Mentoring Alliance

Peer mentoring is one particular form of mentoring that would seem ideally suited to senior women chemists situated at geographically distinct liberal arts institutions. The more egalitarian atmosphere of a peer mentor group with members of similar professional rank is a welcoming venue to share career information and provide support and feedback. Studies have shown that peer mentor groups can be particularly empowering as each member is serving as both a giver and receiver of information (*18–21*). The varied career experiences and achievements of a cohort of women faculty who have reached the senior ranks at their institutions provide a rich resource to utilize for guidance and recommendations.

In 2004 five senior women chemists (four are authors of this chapter) were successful in receiving funding for an interinstitutional initiative supported by a Mellon Foundation faculty development award to a cohort of eight liberal arts colleges (*22*). In this project, "Advancing the Careers of Senior Women Chemistry Faculty through a Horizontal Peer Network", we established a networking peer support group for the purpose of exploring and defining future career aspirations. By meeting together to discuss career goals and establish steps for achieving these aims, the group sought to provide each other with support, advice, ideas, and contacts. We represented the only senior women chemists at our institutions and in the eight-institution cluster (*23*) and, in some cases, the only female chemists at any rank in our departments. In most instances we were the first women faculty hired in our departments with few, if any, female role models in our institutions as we progressed through tenure and promotion to full professor. Thus peer mentoring involving external mentors was a logical approach for our group of women faculty.

To determine the essential components for a successful horizontal peer mentoring approach, we examined the factors that lead to successful and sustained transformations in organizations. We surmised that face-to-face meetings of our small number of participants from distinct but similar institutions over a sustained period of time were central to the success of our initiative. There is ample evidence that, for organizations to initiate and sustain change, members must have a shared vision, use a systems approach that recognizes the interrelationships among participants, and learn as a team particularly through personal commitments made to each other (*24*). Furthermore, faculty participation from multiple institutions in discourse and activities focused on faculty development can lead to more creative approaches and certainly mitigates the sense of professional isolation in pursuing new initiatives. The collaborative team model can invoke a stronger commitment to the goals and a greater appreciation of the dedication of colleagues to the long-range objectives. Research further indicates that successful faculty professional development requires mechanisms that are iterative, systemic, and involve ongoing interactions and interventions (25, 26). The multi-day gatherings for our horizontal mentoring alliance allowed for the personal interactions that are necessary to foster the formation of a support network. With a support network in place and with opportunities to meet regularly, the most lasting career development is likely to occur (27, 28).

Our horizontal peer alliance corresponded electronically for several months prior to our first face-to-face meeting to explore common professional objectives and establish priorities for the first meeting. Electronic communication continued between meetings to continue conversations on issues raised at gatherings, share new information and advice as new professional challenges and achievements occurred, and design the agenda of future gatherings. Prior to the first meeting the group also began the compilation of an annotated bibliography of journal articles and books on career development issues for senior academicians. Reviews of these publications were shared at gatherings and through electronic communications. To facilitate the discussion, members of the group selected several texts to read in common, depending on the particular professional interests of the individuals.

Four meetings of the mentoring group were held from 2004 to 2006 with Mellon funding, with partial gatherings at two professional conferences. At the first meeting, members formulated short-range individual career goals to address over the subsequent months. The group also decided at that meeting to seek the guidance of a career development consultant to enhance their leadership and self-presentation skills. The second meeting was held with an experienced career development and executive coach for women academic physicians and scientists. We addressed a broad range of issues in our coaching sessions including effective communication and relationship-building strategies, self-presentation and self-promotion techniques, and values-based goal-setting. At the third meeting we considered avenues for extending this network to other senior women in chemistry at liberal arts colleges similar to our own institutions and for assisting junior women faculty at our own institutions in their career development. A fourth meeting focused on assessing the impact of this career enhancement project.

Assessment efforts demonstrated this mentoring strategy to be a resounding success. One participant articulated the personal impact of the project on her career as follows: "This grant provided me with the means to meet with and discuss my situation with four other successful and talented women who each had to face their own set of personal hurdles in their career paths. I have been inspired by them and have come to rely on their expertise and decision-making skills to help me in making choices and decisions. … My confidence and self-esteem have soared." Another expressed the value of the experience for both herself and her institution: "The horizontal mentoring network that we have set up will continue to be of great value to me. I know that if I need advice from someone more removed from my setting I now have four individuals, each with different talents, who I can contact for guidance. Knowing the value of this mentoring has reinforced my willingness to provide guidance for junior colleagues as they progress through their careers…. In addition I now realize that the senior women in science at [my institution] need to spend more time together. Each of us is fairly isolated in our own department. Having experienced the utility of a mentoring network first hand I now know that we need the equivalent of this to happen on my home campus. "

An NSF-ADVANCE-PAID Project To Expand the Horizontal Peer Network

Given the success of our initial peer mentoring group, we applied and received funding from the NSF ADVANCE PAID (Partnerships for Adaptation and Implementation) program in 2006 to continue our mentoring approach. Our project established a network of four five-member "horizontal" mentoring alliances of senior women scientists at private liberal arts institutions – three alliances were composed of chemists at the full professor rank and the fourth was composed of physicists at the full professor rank. A significant amount of research occurred to survey the composition of chemistry and physics departments at over 200 liberal arts colleges around the country. The members of the alliances were selected on the basis of their existence as the lone senior female faculty member in their department and often for their presence as the singular female faculty member in their department. Geographical diversity was a key objective in three of the alliances in an effort to bring together women whose institutions might not already be participating in a regional consortium. We did recognize, however, that travel times could hinder meeting during the academic year, so a more regional association of alliance members was sought in one alliance to test the impact of reducing that constraint. No effort was made to match women in similar subdisciplines of chemistry or physics. With one alliance the selection of women who had fairly recently attained full professor status was the aim. We also deliberately chose women from twenty different institutions to maximize the impact of campus outreach efforts.

Each alliance was free to determine their own meeting times and locations. Some alliances chose to meet on each other's campus to benefit from learning about each institution. Other alliances held gatherings in conjunction with professional society conferences in order to minimize travel. Still others, when time was tight during the academic year, chose hub cities and even airport hotels for convenient air travel and maximum time for interaction. At the first gathering of each alliance, in addition to getting to know one another, each alliance participant shared her individual short- and long-range career goals and the alliance decided on areas of career development to address as a group (e.g., leadership, self-presentation and self-promotion skills, dealing with difficult colleagues, effective communication and relationship-building strategies, etc.). Funds were available for external consultants to provide guidance in these areas and for the purchase of reference books as additional resources. Each alliance chose its own means of electronic communication and/or teleconferencing to stay connected between gatherings and to further promote the sharing of advice, ideas, and contacts. Significant numbers of the members of the chemistry alliances gathered for dinner at national meetings of the American Chemical Society to create an expanded network of colleagues and share news about the activities of each alliance. Finally, a variety of outreach activities on home campuses extended the impact of the career development expertise attained by project participants. These outreach activities included, for example, book discussion groups on faculty development topics for the women science faculty on a given campus; sponsorship of a consultant visit to conduct a strategic career planning workshop for the women science faculty with individual career planning meetings and sessions on negotiation and brainstorming on critical career issues; and a visit to campus of an external speaker to provide professional development for all science faculty (male and female) in the form of information about significant contemporary interdisciplinary research questions and career paths and internship opportunities for students. As a culminating event of the project, a summit meeting was held in Washington, D.C. for all project participants and more than thirty additional senior chemistry and physics female faculty members at liberal arts colleges to identify and create resources that address career development issues for senior women at liberal arts institutions and disseminate best practices on horizontal mentoring strategies for academic women.

The Efficacy of the Horizontal Peer Mentoring Approach

Our project to establish horizontal peer networks of senior women chemists and physicists at private liberal arts institutions has the following distinguishing features to insure effectiveness:

- an approach focused on senior women chemists or physicists who are employed in the distinctive environment of a private liberal arts campus,
- a structure that enables multi-day gatherings that foster the personal interactions necessary to form a committed cohort of faculty to serve as peer mentors,
- mechanisms for regular follow-up to maintain the support network and mitigate professional isolation, and
- professional development activities tailored to the specific needs of the participants and designed to enable these senior women scientists to serve as effective leaders of institutional change on their own campuses and in their professional associations.

Our project evaluator (and an author of this chapter) conducted an ethnographic study using qualitative research methods, and her results show this form of peer mentoring to be particularly effective. Our summative evaluation is still in progress, but formative evaluation involving interviews with project participants explored early outcomes of participation in the initiative and revealed many benefits of the mentoring approach. Alliance members were asked their views about the efficacy and relevancy of the structural model (i.e., a horizontal mentoring alliance) in practice, their thoughts about barriers and supports to using this model, its sustainability, and where it might be usefully replicable.

Participants overwhelmingly agree that the alliances promote the sharing of ideas, experiences, and expertise. Furthermore, the composition of the alliances with members from different institutions was valued in that it provided a different perspective from an outsider's point of view and an opportunity to be open and honest without fear of competition or reprisal. Nearly 75% of participants agreed that being part of a horizontal mentoring alliance had given them more confidence to "speak up for myself," ask for what they wanted from their departments, accept due recognition for their professional work and contributions and permission to focus more time and attention on their professional goals. Three-quarters of participants also noted that, aside from strong professional support, they had developed friendships with other alliance members that would last beyond the life of the grant and that a benefit of participating was simply in talking and socializing with other women having similar career paths and interests The career development focus of alliance meetings, network gatherings, and horizontal mentoring activities also contributed to many major professional developments for the project participants. A key aspect of the initiative is the articulation of short- and long-range career goals by each participant and the formulation of action plans to attain the stated professional goals. An extensive array of enhanced leadership and career opportunities have resulted including endowed professorships, institutional awards for teaching and service, invited lectureships, and offices in professional organizations. Participants also noted a range of additional benefits that included the transfer of gains back to their own institutions in terms of a renewed effort to mentor women and in terms of interactions with deans and other institutional administrators who were actively interested in the horizontal mentoring alliance initiative and were interested in seeking ways to support women science faculty. Coding of the interview data also revealed comments focused on the alliance meetings, the topic of mentoring, geographical issues associated with career development and alliance functioning, professional development issues associated with differences between R1 and liberal arts college settings, replicability and sustainability of the horizontal mentoring alliance, career satisfaction, comments on whom the horizontal mentoring alliance best serves, as well as additional gender, departmental, and institutional issues. These comments helped to structure subsequent alliance gatherings and develop communication among the alliances. Overall, baseline outcomes from the external evaluation demonstrated that the rationales underlying the development of the Horizontal Mentoring Alliance initiative were accurate. In practice, the mentoring model that was implemented was found to be highly effective in addressing issues particular to senior women faculty members in the sciences at liberal arts colleges, and thus, successful in achieving project goals of reducing members' isolation, increasing their access to professional networks and advice, and in promoting their career advancement. It is notable that qualitative findings from participant interviews align with the varied research and discussions in the literature concerning women science faculty and academe. The strong benefits to alliance members, their colleagues and institutions suggest that effective mentoring is needed and beneficial at all levels of one's academic career. As a model, horizontal mentoring might well be adopted by others seeking to effectively promote women science faculty members' advancement in academe.

Conclusion

The formation of these horizontal mentoring alliances has had significant direct impact on the career development of the twenty senior women participants and additionally developed a cohort of leaders of institutional change at the participants' home institutions. Participants cite a range of personal benefits from involvement in this initiative including opportunities to network with senior women science faculty in liberal arts institutions; time to engage in career development discussions aimed at enhancing leadership, visibility, and recognition on campus and in the broader academic community; and occasions to develop mentoring paradigms that can be used with students, junior female faculty colleagues, and other senior female faculty colleagues. This horizontal mentoring strategy has also enabled participants to realize numerous individual gains that have impacted both their professional and personal lives. It is our belief that, for senior women faculty seeking new avenues of career development resources, a horizontal mentoring approach might indeed offer a viable mechanism .

This material is based upon work supported by the National Science Foundation under Grants NSF-HRD-0618940, 0619027, 0619052, and 0619150. Any opinions, findings, and conclusions or recommendations expressed in this material are those of the author(s) and do not necessarily reflect the views of the National Science Foundation.

References

1. ADVANCE: Increasing the Participation and Advancement of Women in Academic Science and Engineering Careers Program Solicitation; NSF 05-584; National Science Foundation: Arlington, VA, 2005.
2. Science and Engineering (S&E) doctoral holders employed in unversities and 4-year colleges, by broad occupation, sex, years since doctorate, and faculty rank: 2002, Table H-22. NSF 09-305. National Science Foundation, Division of Science Resources Statistics, Survey of Doctorate Recipients. http://www.nsf.gov/statistics/wmpd/pdf/tabh-22.pdf (accessed August 19, 2010.
3. Ragins, B.; Cotton, J. Mentor functions and outcomes: A comparison of men and women in formal and informal mentoring relationships. *J. Appl. Psychol.* **1999**, *84*, 529–550.
4. Ambrose, S. *Journeys of Women in Science and Engineering*; Temple University Press: Philadelphia, PA, 1997.
5. Kanter, R. *Men and Women of the Corporation*; Basic Books, New York, 1977.
6. Burke, R. Mentors in organizations. *Group Organ. Stud.: Int. J.* **1984**, *9*, 353–372.
7. Ragins, B.; Cotton, J. Mentor functions and outcomes: A comparison of men and women in formal and informal mentoring relationships. *J. Appl. Psychol.* **1999**, *84*, 529–550.

8. Cawyer, C.; Simonds, C.; Davis, S. Mentoring to facilitate socialization: The case of the new faculty member. *Int. J. Qual. Stud. Educ.* **2002**, *15*, 225–242.

9. Boice, R. Lessons Learned about Mentoring. In *New Directions for Teaching and Learning*, No. 50; Sorcinelli, M., Austin, A., Eds.; Jossey Bass: San Francisco, 1992; pp 51–61.

10. Luna, G.; Cullen, D. L. Empowering the Faculty: Mentoring Redirected and Renewed. ASHE-ERIC Higher Education Reports No. 3; 1995.

11. Madison, J.; Huston, C. Faculty-faculty mentoring relationships: An American and Australian perspective. *NASPA J.* **1996**, *33*, 316–330.

12. ADVANCE Portal. http://www.portal.advance.vt.edu/ (accessed August 2010).

13. Packard, B. A 'Composite Mentor' Invention for Women in Science, American Educational Research Association Annual Meeting, Montreal, 1999.

14. Chesler, N. C.; Chesler, M. A. Gender-informed mentoring strategies for women engineering scholars: On establishing a caring community. *J. Eng. Educ.* **2002**, *91*, 49–56.

15. Simeone, A. *Academic Women: Working towards Equality*; Bergin & Garvey Publishers, Inc.: Boston, 1987; p 101.

16. Women Chemists 2000. American Chemical Society. http://www.chemistry.org/portal/resources/ACS/ACSContent/careers/empres/WCC_contents.pdf (accessed January 5, 2006).

17. 2005 Carnegie Classification of Institutions of Higher Education. The Chronicle of Higher Education Almanac. http://chronicle.com/article/2005-Carnegie-Classification/47991/ (accessed August 19, 2010).

18. Shapiro, E.; Haseltine, F.; Rowe, M. Moving up: Role models, mentors, and the 'Patron System'. *Sloan Manage. Rev.* **1978**, *5*, 51–58.

19. Clark, S.; Corcoran, M. Perspectives on the professional socialization of women faculty: A case of accumulative disadvantage? *J. Higher Educ.* **1986**, *57*, 399–414.

20. Johnsrud, L. K. Mentor relationships: Those that help and those that hinder. *New Direc. Higher Educ.* **1990**, *18*, 57–66.

21. Chesler, N. C.; Single, P. B.; Mikic, B. On belay: Peer-mentoring and adventure education for women faculty in engineering. *J. Eng. Educ.* **2003**, *92*, 257–262.

22. Andrew W. Mellon Foundation Grant for Faculty Career Enhancement. List of Awards. Interinstitutional Initiatives, December 2003. http://www.depauw.edu/admin/acadaffairs/facdev/careerenhanceclusterawards2003.asp (accessed January 3, 2006).

23. The eight-institution Mellon cluster consisted of Denison University, DePauw University, Furman University, Harvey Mudd College, Middlebury College, Rhodes College, Scripps College, and Vassar College.

24. Senge, P. M.; Kleiner, A.; Roberts, C.; Ross, R. B.; Smith, B. J. *The Fifth Discipline Fieldbook: Strategies and Tools for Building a Learning Organization*; Doubleday: New York, 1994.

25. Guskey, T. *Evaluating Professional Development*; Corwin Press: Thousand Oaks, CA, 2000; pp 14–30.

26. Collins, A.; Brown, J.; Newman, S. Cognitive Apprenticeship: Teaching the Crafts of Reading, Writing, and Mathematics. In *Knowing, Learning, and Instruction: Essays in Honor of Robert Klaser*; Resnick, L., Ed.; Lawrence Erlbaum: Hillsdale, NJ, 1989.

27. Sunal, D. W.; Hodges, J.; Sunal, C. S.; Whitaker, K. W.; Freeman, L. M.; Edwards, L.; Johnston, R. A.; Odell, M. Teaching science in higher education: Faculty professional development and barriers to change. *Sch. Sci. Math.* **2001**, *101*, 246–257.

28. Herr, K. Exploring excellence in teaching: It can be done! *J. Staff, Program, Organ. Dev.* **1988**, *6*, 17–20.

Chapter 11

Promoting Mentoring among and for Women in Chemistry: The Experiences of COACh

Jean Stockard,[1] Jessica Greene,[1] Priscilla Lewis,[2] and Geraldine Richmond[*,3]

[1]Department of Planning, Public Policy and Management, University of Oregon, Eugene, Oregon 97403
[2]COACh Program, University of Oregon, Eugene, Oregon 97403
[3]Department of Chemistry, University of Oregon, Eugene, Oregon 97403
[*]E-mail: Richmond@uoregon.edu. Phone: (541) 346-4635.

Mentoring is considered an important factor for why women are underrepresented in academic science and engineering departments. COACh, the Committee on the Advancement of Women Chemists, has been working on programs involving mentoring of women scientists for the past decade. COACh has sponsored numerous career-oriented workshops for academic women chemists and has been instrumental in developing workshops for department heads that all have some component of mentoring built into them. This paper uses data gathered by COACh at COACh workshops that examine women chemists' mentorship experiences. Through a series of comments gathered from these women chemists, insights can be gained on issues such as what mentors have been effective in their lives, what mentors do, the effectiveness of formal mentoring programs, the changing mentor/mentee role over the course of a career, why mentoring often doesn't happen and what factors can contribute to having a positive mentoring experience. The article ends with a discussion of ways that COACh has promoted mentoring and the apparent results of these efforts. It ends with a brief discussion of future research that needs to be done in this area and lessons for policy and action.

© 2010 American Chemical Society

A growing body of research documents numerous factors that contribute to women's under-representation in the top tiers of technical and scientific fields. This literature describes the ways in which biases against women, particularly those that are more subtle and implicit, can translate into lower salaries, slower rates of promotion, and less recognition through honors and awards for women relative to their male colleagues (1–3). In 1999 a small group of senior women chemistry faculty from around the United States began meeting to discuss their concerns that women in their field were not experiencing the same career opportunities and advancement patterns as men. With seed funding from the Camille and Henry Dreyfus Foundation, they formed an organization called COACh, the Committee on the Advancement of Women Chemists.

Over the last decade COACh has sponsored numerous career-oriented workshops for academic women chemists and has been instrumental in developing workshops for department heads. One aspect of this work is encouraging mentorship relationships for and by women, an area recognized by the COACh founders as important for career advancement. Mentoring, or, more specifically, the lack there of, is considered an important factor for why women are underrepresented in academic science and engineering departments (4). A recent review of the literature on mentoring in academic medicine found that having a mentor was associated with greater research productivity and higher likelihood of receipt of federal grants, as well as higher job satisfaction and confidence (5).

This paper uses data gathered by COACh to examine women chemists' mentorship experiences and then discusses ways that COACh has promoted mentoring and the apparent results of these efforts. We end with a brief discussion of future research that needs to be done in this area and lessons for policy and action.

The Mentorship Process

Before attending COACh sponsored workshops, which have been held in conjunction with national professional meetings since 2001, the women participants completed surveys that included questions about their mentorship experiences. In addition, a sub-sample participated in in-depth interviews, lasting up to one hour, regarding their careers. For this paper, we examined data from 255 surveys and 47 interviews. The data were gathered over a span of several years and represent women chemists at various stages of their careers. Thus, they provide information of a broad cross-section of women in the field. The sections below examine who their mentors were and the type of help they received, how mentorship roles sometimes changed over the course of a career, women's experiences with formal mentorship programs, views about why mentoring does not happen more often, and examples of individuals and departments that have had successful mentoring experiences.

Who Mentors?

Slightly more than a third of the women (38%) reported having a mentor during their education and professional training, and somewhat more (44%) reported having a mentor during the first 10 years of their career. As would be expected, mentors almost always held positions that were more senior than the women and had more experience in the field. (Exceptions were those who reported help from a spouse or graduate student and/or post-doctoral colleagues who were just a few years senior.)

A large proportion of the women who had mentors during their training mentioned the influence of their teachers from middle-school teachers who interested them in science to undergraduate instructors, to those with whom they worked in graduate and post-doctoral programs:

"When I was in high school, I think my high school teacher was really, really good...he was an inspiration and that was the most important role that he played."

"My high school chemistry teacher was very supportive and he really wanted me to get my degree in chemistry. He was very encouraging....I still correspond with him...At [undergraduate school] there was a woman faculty member ... [who] took a liking to me and she was very encouraging...The other faculty ...were [also] incredibly wonderful and mentoring."

"My undergraduate research advisor was very, very encouraging...My post-doc advisors were good. My PhD advisor and I would talk about all sorts of things, career-related."

Mentors to women early in their academic careers were often senior colleagues. Several women mentioned their department chairs and, occasionally, division deans as providing support and advice. Senior colleagues mentioned were often chemists, but mentorships also crossed disciplines. In general, respondents indicated that the people who were most likely to provide help were those who were most comfortable in their own careers. As one person put it,

"I have found that people who are very senior and very accomplished, who are basically secure with themselves, help the younger ones. Those people are great."

Although the women chemists reported having both male and female mentors and we do not have exact counts by gender, the representation of women as mentors appeared to be larger than their representation in the field as a whole. In other words, women reported often looking for other women to serve as a mentor. In addition, women of color sometimes reported choosing mentors that spanned both gender and race/ethnicity. For some, the racial/ethnic similarity was especially important. An Asian-American woman described:

"I think one of the good things about being in both ethnic and gender minorities in this field is that I can draw from both. They have similarities. I find both the Asian and the women's community are very helpful because they realize the issues facing me."

Similarly, an African American woman noted,

"I have interacted with the black faculty in the college of [....], all men, and we just connect and click, spiritually and just fundamentally we're all kind of on the same page....They respect me."

What Do Mentors Do?

In responses to open-ended questions on the surveys and lengthy descriptions in the personal interviews, the women described how mentors had helped their careers. Most often mentioned was the way in which mentors provided career advice, from assistance in choosing research topics and graduate schools, to feedback on research proposals and writing, to advice on where to apply for jobs and strategies to use in the job search. For example,

"They reviewed papers and proposals, gave me advice for different types of decisions, encouraged me to stick my neck out and take some chances."
"She helped me through the application process....She gave me tips on interviewing and negotiating."
"Although he was in a very different field, he [my mentor] would read grants and I would talk to him about how to run my group and how my research was going. My group held joint group meetings with his group regularly and he would serve on my students' committees."
"[A senior faculty member and I] taught different sections of the same course and he initially provided the syllabus, set up all the demos, and let me come to his class to observe."

A number of women also described how, in addition to providing specific advice, their mentors went out of their way to promote their careers. This involved actions as diverse as introductions to prominent researchers in their area, invitations to speak at professional meetings, talking about their work to funders and others who could promote their career, and nominating the women for awards. For example,

"He put forth my name for editing journal articles, grant reviewing, and conference organizing. He encourages me to publish and write grants and connected me to funding opportunities and collaborations."
"One...knew my NSF program chair personally, and got me to call up my program chair and introduce myself and get some general grant-writing advice that was very useful....My thesis advisor also acted as a mentor after I started my faculty position, by inviting me to speak at a National ACS meeting very early in my career."

Finally, several of the women mentioned how mentors provided emotional support and encouragement. They helped build their confidence, they provided advice in times of stress and conflict, and they served as "cheerleaders" as career transitions were made. They also made social overtures that promoted feelings of belonging and inclusion. For example,

> "He taught me not only the research stuff, but also to value myself as a strong individual with meaningful contributions to make."
> "They often provided assistance with networking and dealing with political situations both within the department and institution."
> "She made sure that I got whatever information I needed, made sure that I joined her group for lunch, [and] did all the right stuff."

A woman who dealt with issues related to both gender and race-ethnicity reported how her mentor helped her deal with discrimination based on both characteristics:

> "He made me become aware of how to deal with people, how to be more vocal, how to talk to people, and so on and so forth, and kind of coached me...I really appreciate him. He really spent time with graduate students, so I've been very lucky."

Changing Mentor/Mentee Roles during a Career

Several women reported changes in the role of the mentor as their careers progressed. Women who were more senior reported seeking out people at their own career stage for advice and counsel, often in other fields. However, several of those at more advanced stages of their careers noted the difficulty of finding someone who felt comfortable working with them at that stage and even that a once productive and rewarding mentor-mentee relationship had soured due to professional conflicts and other issues. For example,

> "He changed his mind [about mentoring me] last year and decided that I was intimidating and has stopped mentoring me."
> "He decided that I was intimidating to the staff and stopped mentoring me and chooses instead to block my advancement and full participation in the department....I believe that they are intimidated because I am a smart, confident, capable woman, and they don't want to be led by me. The trouble started when I assumed a leadership position that involved staff reporting to me."
> "My chair kind of protected me and respected me, as long as I did stuff for him. Later on, when I wanted to challenge him, he didn't like it."

On the other hand, some of the interviewees reported long-lasting relationships that matured and grew, remaining helpful and highly regarded for many years.

"My boss and another colleague in my Department have been excellent mentors. They have been wonderful role models who have created endless valuable professional opportunities for me. They have included me in projects, sent opportunities my way, nominated me for awards, given advice, reviewed my work, given excellent feedback….everything one would expect from a mentor. They have made a significant impact on my career, and as a result of working with them I believe I have been probably 4 times more productive than I would have without their guidance and leadership."

"I have continued to stay in touch with my PhD advisor and also my other colleagues along the way to talk about the ups and downs of jobs, applications, etc."

"My former research advisor continues to be a wonderful mentor. While many of the things he did for me in early years are no longer necessary, he still helps with things like award nominations and his words of admiration and approval still make me feel great."

Formal Mentorship Programs

Some respondents reported that their departments had official or organized mentorship programs. Occasionally the respondents reported that the officially appointed mentor was helpful. For example, one noted that

"[my] assigned mentor from the department…stood up for me during hard times [and] listened…[This was] very important to me because the tenured women and some of the "older male" faculty in my department were not so supportive."

Several others, however, noted difficulties with the mentor that had been assigned to them and that the people who ended up being most helpful to them were often not those who had been officially designated as mentors. Because the impact of formal mentoring programs was not an explicit focus of the interviews or survey questions, these examples are especially telling.

"I was assigned to work with someone who had no interest and wasn't willing to take the time to mentor."

"The mentors assigned to me by my department did not wish to be mentors."

"[I] enrolled in a mentor program at my university for new faculty, but my assigned mentor had a very busy year and ended up backing out of the program"

"I had one [a mentor] "assigned" by the Dean, but this was useless. I developed an external female mentor in another department…This current mentor helps identify potential awards, answers questions about career planning (which committees to serve on and which not to, etc.) She has also been active in helping to bring the issue of lack of women to the Dean's attention."

"A mentor was formally assigned by the department. But I don't trust him. He has already tried to manipulate me at least twice. Actually the first time that I talked to him after I was hired, he told me that he had argued against hiring me, though I will occasionally ask him for advice."

"I didn't find [the formal mentoring committee system] very useful because one of the people that I sort of had a conflict with … was my main mentor."

"All that [my formal mentor has] done basically is told me that I need to get external funding, and that that's what I need to focus on, writing big grants rather than small grants….he's even being paid to be my mentor, and…I don't feel like he's put any time into that….they actually have these formal guidelines…for how it's supposed to be done, and he didn't really have any interest because it's just too time consuming."

"When I went to … they teamed us all up with mentors, and we met each other for a few times. My mentor was a very nice man but he just didn't have time after we had all been assigned. It just didn't work out…You have to find out who's interested in what type of research so you can find out whom you can collaborate with quickly."

Why Does Mentoring Not Happen?

Recall that over half of the women surveyed by COACh did not have mentors during their educational training and early careers. Some of these women reflected upon why they did not have mentors. Some of their responses involved their own actions, such as not knowing how a mentor might help and/or not knowing how to approach others for help. For example,

"I did not know [I] needed one and none were available."

"I did not understand the system, was unaware of mentoring."

"No one ever suggested I should have a mentor."

"I am currently in my fourth year as an assistant professor and I have had no mentors thus far. I was the first female faculty member hired in my department in a new research area. Therefore, I did not have many opportunities to interact with other faculty, unless initiated by me. I often feel awkward asking questions that might make me look "stupid," and therefore I often feel isolated in my department."

Others (and occasionally the same respondents) noted that mentorship of young faculty did not seem to be a priority at their institution and/or that more senior faculty members were not interested in pursuing this role. For example,

"It was difficult [after becoming a faculty member] to find a formal mentor who had time or who was genuinely interested in investing in you. I have an official mentor from the university who I would contact when I had questions or needed feedback. I would be the person to initiate this since this person was also the mentor for several other people in the department and he was also a very busy person….The natural mentor in my field did not seem interested and would often say that they did not have time to meet with me."

"[There were] no women in senior positions and no men who seemed willing to help."

"There was no one willing to dedicate the time to mentoring during my early tenure track years."

"There is no requirement to have a mentor to assist junior faculty in the career development in our college."

"There was no mechanism for having a mentor [in my department]."

Making Mentoring Happen

Finally, however, the surveys and interviews provided insight into how mentoring can be promoted and developed. In part, this involves individuals seeking out mentoring opportunities. One person described her experiences,

"…Starting in 1984, I've gone to every single ACS meeting. There are a lot of people who do that, but there are also a lot of people who don't do that. I did with the express purpose to really get to know people. And have them get to know me. I presented a lot. I went to the ….conferences and was generally very impressed how easy it was to get to know people, to talk to them and sort of talk myself up. That was a little difficult.…."

Others described how their departmental culture and climate made mentoring an integral part of the organization's day-to-day life and norms of collegiality.

"Even though there was no assigned mentor to me in the department at […], I think the environment mentored me and everybody else in terms of group development, the whole package."

"Our department is really good at making sure that there are collaborations. Like the senior faculty would be like I want you to help me with this grant.. And we still do that when someone is new there, we try to get them started [on] something that is collaborative. So, [there are] not true mentors I would say, but when someone is new we try to get them started with collaborating."

In short, these examples illustrate how some women have promoted their own mentoring activities and how departments can develop cultures and practices that support and mentor all faculty members. The activities of COACh have incorporated these themes, and we now turn to a description of these activities and a discussion of their outcomes.

How COACh Promotes Mentoring

COACh has used two different routes to alter the culture of mentoring in Chemistry. One focuses on women chemists and the other focuses on departmental cultures and leadership.

COACh's professional development workshops for women faculty have been held at national professional meetings and at individual institutions. Workshops have sometimes focused on participants at different points in their careers, such as post-doctoral fellows, assistant professors, or more senior faculty; and on participants with multiple concerns and possibilities of bias, such as women who are racial-ethnic minorities. Since 2001, over 400 women chemistry faculty members have attended these COACh developed workshops at national professional meetings. It is estimated that over one third of the women faculty who hold tenure track positions at the top 100 chemistry departments in the country have undergone COACh training at the national meetings or at their home institutions.

The workshops have two general aims: first, they are designed to help women develop skills to facilitate their career progress. Experts are employed to provide training in areas as varied as communication tactics, negotiation skills, and leadership strategies. Second, they are designed to provide a venue for participants to network with other successful women chemists. By bringing women together from around the country in an atmosphere that is professional, but removed from immediate work obligations and roles, they promote the development of network ties and establishing new relationships. They explicitly seek to broaden women's networks by hosting social events after the workshops.

Evaluations of the workshops, both immediately after the events and up to several years after their occurrence indicate that participants believe that the trainings have contributed significantly to their career progress and their professional well-being. The results also indicate that the workshops motivated the women to help others within their profession. For instance, in response to a survey sent to COACh attendees several years after their first participation, two thirds of the women said that the skills they had learned through COACh had helped them develop supportive networks, either quite a lot (22%) or a fair amount (45%). Only six percent of the women said the skills had not helped them in this area. Almost as many women (60%) said that the skills acquired through COACH had helped them mentor others.

The data also suggest changes over time in the percentage of women reporting that they have received mentoring. Women who have attended COACh workshops in recent years and those who received their PhDs more recently have been significantly more likely than other women to report mentoring both during their education and the early stages of their careers (6).

Changing Department Climates

COACh has also focused on academic leaders within chemistry and, in 2006, helped conduct a workshop entitled "Building Strong Academic Chemistry Departments through Gender Equity." The workshop was sponsored by three federal agencies that provide the vast majority of research funding to chemists in academic Department heads from the departments that receive the most federal research and development money and/or produce the largest number of PhD students were invited to attend. The workshop was designed to develop awareness of the problem of women's under-representation in academic chemistry, to motivate leaders in the field to work for change, to develop concrete steps to address the inequities, and to obtain commitment from participants to promote changes in their home departments. As with other COACh activities, the efficacy of this workshop was evaluated. Attendees were required to complete a questionnaire that examined their attitudes and perceptions regarding women's representation in chemistry both before attending the workshop and after returning to their home institutions. In the months after the workshop the chairs were also charged with reporting on specific actions they had taken with regard to gender equity on a password protected website open to other department chairs.

Analyses of these data indicated that the heads who attended the workshops became more aware of how the lack of mentoring can impact women's career progress and committed to changing their departments' culture and ways in which it supports young faculty. For instance, before the workshop, half of the attendees believed that few available mentors was either "not an issue" or "not important" in affecting women's career progress. After attending the workshop, however, only a quarter of the participants held these views, a change that was highly statistically significant. The chairs' reports of the action items to which they committed after attending the workshop also indicated a strong focus on mentoring. Of the 45 heads who listed goals and action items on the group's website, over half (26) indicated activities that would assist career development, such as a mentoring program. (See Stockard et al, 2008 (7) for a full report on the workshop.)

Looking to the Future

There are, of course, limitations to the data reported above. For instance, even though the women in our sample represent a relatively large proportion of those in the discipline, we have no way of knowing how representative they are of the total group. In addition, we do not have comparable data on men or on people in other disciplines. Finally, we are not able to provide controls that can help indicate the extent to which the results we found are attributed to the interventions of COACh or to other changes within the discipline or academic enterprises as a whole. Thus, future research in this area is necessary.

That said, we believe that the results summarized here provide important insights that can guide those concerned with promoting successful careers for all scientists and especially those, such as women and members of other under-represented groups, who have historically faced bias and discrimination. First, our results indicate that mentors are important. Most of the respondents believed that their mentors provided important career assistance. Second, only a minority of the women in our sample reported having mentors during their education and professional training or the first ten years of their career. Thus, there may well be an unmet need for mentoring help. Third, the results suggest that mentoring may have become more common in recent years, perhaps resulting from women seeking out and providing more mentoring support as well as department heads becoming more aware of this need. The evidence associated with the evaluations of COACh sponsored workshops suggest that these efforts may have contributed to these changes. Fourth, official mentoring programs are not always effective and it appears that care must be taken to ensure that mentors and mentees are well matched and that mentors are committed to their roles. Even more important may be developing departmental and discipline-wide cultures that are supportive and mentoring to all young scholars, no matter what their background may be.

References

1. Committee on Science Beyond Bias and Barriers. *Fulfilling the Potential of Women in Academic Science and Engineering*; The National Academies Press: Washington, DC, 2007.
2. Greene, J.; Lewis, P.; Richmond, G. L.; Stockard, J. *J. Chem. Ed.* **2010**, *87*, 381–385.
3. Valian, V. *Why So Slow? The Advancement of Women*; MIT Press: Cambridge, MA, 1999.
4. Chandler, C. *NWSA J.* **1996**, *8* (3), 79–10.
5. Sambunjak, D.; Straus, S. E.; Marusic, A. *JAMA, J. Am. Med. Assoc.* **2006**, *296*, 1103–1115.
6. Greene, J.; Lewis, P.; Richmond, G. L.; Stockard, J. *J. Chem. Ed.* **2010**, *87*, 386–391.
7. Stockard, J.; Greene, J.; Lewis, P.; Richmond, G. *J. Women Minorities Sci. Eng.* **2008**, *4*, 1–27.

Recommendations for Individuals

Chapter 12

Integrating Work and a Personal Life: Aspects of Time and Stress Management for Senior Women Science Faculty

Julie T. Millard[*,1] and Nancy S. Mills[2]

[1]Department of Chemistry, Colby College, Waterville ME 04901
[2]Department of Chemistry, Trinity University, San Antonio TX 78212
*jtmillar@colby.edu

One of the many myths surrounding college teaching is that professors are only working when they are actually in class. Particularly for science faculty at liberal arts colleges, time in the classroom is only a small fraction of a typical workday. Research, mentoring, and service are just some of the other demands on our time. Adding personal responsibilities increases time pressure, which can lead to a constant battle with stress. Increased stress in turn reduces productivity and leads to a decline in physical and emotional well-being. In this paper, we address some of the consequences of the stress encountered in an academic career, some of the special challenges for women science faculty, and some strategies for achieving a better balance between the professional and personal.

Too Much To Do

"There is never enough time, unless you're serving it."
-Malcom Forbes

A constant feeling of having too much to do and not enough time to do it can lead to a high degree of work-related stress. This phenomenon is certainly not isolated to academics but is shared by many working people. A 2007 Gallup poll found that 47% of Americans report that they do not have enough spare time (1). Most people also report that they experience stress in their daily life, with a correlation between having insufficient spare time and experiencing stress (Figure

© 2010 American Chemical Society

1). For example, 54% of Americans who reported having insufficient time were frequently stressed. On the other hand, only 27% of Americans who reported having sufficient spare time said that they frequently experienced stress.

Hidden Consequences

What are the costs for professional women who don't have enough time for a personal life? Few can forget the 1986 *Newsweek* article that claimed that a college-educated single woman over the age of 40 was more likely to be killed by a terrorist than to get married (2). This article referred to a study based on figures obtained from the Census Bureau (3), yet one of the study's coauthors said that *Newsweek* took the findings out of context (4). For example, the study did not differentiate between women who wanted to get married and those who didn't for any number of reasons. Furthermore, the terrorist angle was mere hyperbole on the part of the author, yet it struck a chilling note with many women of the time. Twenty years later, the numbers looked quite different, with about 90% of baby-boomers of both sexes either married or expected to marry (5). Nonetheless, the 1985 study highlighted a growing trend on the part of professional women to delay marriage until their educations were complete and their careers established.

Over the past few decades, the media has also bombarded women with the message that to delay childbirth may result in infertility. However, for women who choose an academic career, the tenure track often conflicts with the ideal reproductive years (6), leading many to opt out of childbearing completely. Whereas 33% of high-achieving women are childless at age 40, female academics have the highest professional rate of childlessness at 43% (7). It appears that the perception of academics conflicts with the reality. As said by Hal Cohen (8), "It would seem that a university-- with its ability to allow teachers to work from home, its paid sabbatical semester and its famously liberal thinking-- would be an ideal place to balance career and family. But by all accounts, the intense competition, the long hours and the unspoken expectations of the academy's traditionally male culture conspire to make it really, really hard to have a baby and be a professor."

For professional women who do choose to have children, is there a cost? The data suggest that there is. Correll and coworkers (9) looked for a "motherhood penalty" in the job market. In their studies, two nearly identical application packages were prepared for the same job. The pair of fictional same-gender candidates varied only in the type of community service each had performed, with one applicant easily identifiable as a parent by participation in a parent-teacher organization. Candidates were evaluated both by laboratory subjects and by actual employers. For female applicants, "mommification" of candidates resulted in significant negative penalties, such as fewer callbacks and recommendations for lower starting salaries. In contrast, "daddification" actually benefited male candidates. The authors interpreted the motherhood penalty as arising, at least in part, from the expectation on the part of the evaluators that mothers are more apt to be distracted from their professional obligations by their families. On the

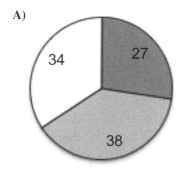 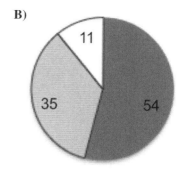

Figure 1. Having insufficient spare time leads to increased stress. A) Frequency of stress among Americans reporting that they had sufficient spare time: ■ *frequently experience stress;* ▨ *sometimes experience stress;* □ *rarely/never experience stress. B) Frequency of stress among Americans reporting that they had insufficient spare time:* ■ *frequently experience stress;* ▨ *sometimes experience stress;* □ *rarely/never experience stress. Data from reference (1).*

other hand, cultural expectations of fathers are not incompatible with the role of "ideal worker."

Bias Avoidance

Given the motherhood penalty in the workforce, it is not surprising that academic women often employ the strategy of "bias avoidance"; that is, engaging in behaviors intended to minimize or hide intrusions of family life on academic commitments (*10*). Furthermore, a growing number of Americans are faced with the challenges of eldercare. Even those who never had children or now have grown children may find themselves responsible for their aging parents. Bias avoidance strategies can also be used to hide leisure activities that could be perceived as distracting from one's professional life, such as training for a marathon or rehearsing for a theatre production. Bias avoidance behaviors can be classified as either productive behaviors, which improve work performance, or nonproductive behaviors, which hinder both work and/or family performance (Table I; (*10*)).

In a study of bias avoidance in the academy, Drago and coworkers surveyed 4,188 college faculty members from the gendered fields of English and chemistry (*10*). Not only are there differences in the percentages of female professors at the college level (in 1999, 60.1% of university English teachers were female compared to 19.5% in chemistry), but the environments of these departments differ considerably, with absences from a chemistry laboratory much more visible than absences from an English office. In general, women were found to engage in several bias avoidance behaviors more frequently than men (Table II).

Differences across disciplines included the facts that women in chemistry reported returning to work too soon after the birth of a new child more often than women in English and men in chemistry were more often partnered and parents than men in English. The atmosphere at the home institution also influenced the degree of bias avoidance behaviors, with supportive supervisors and institutions reducing the frequency. Women at teaching institutions were more likely to miss their children's important events than women at research institutions, presumably because heavier teaching loads lead to decreased flexibility. The authors conclude that greater gender equity and family-friendly policies may reduce the incidence of bias avoidance behaviors.

Increased Demands on Women Faculty in the Sciences

It is clear that science is a demanding field, with a recent study reporting that partnered science faculty at thirteen of the top research universities in the United States work for an average of almost 60 hours per week (*11*). While there was no difference between hours worked professionally by men (56.4 hours) and by women (56.3 hours), the authors did find differences in the amount of housework performed by each sex, with females spending about twice as much time on these tasks as their male peers. Female scientists in opposite-sex partnerships do 54% of core household jobs in their homes, whereas partnered male scientists do just 28%. These tasks, which include grocery shopping, cooking, cleaning, and laundry, require an average of 19.3 hours per week. While males spend more time on average than females on yard and car care, home repair, and finance, these jobs are less time consuming, averaging only about 4.7 hours per week. These findings suggest that on average women scientists have greater demands on their time at home than their male counterparts.

Special Demands on Senior Women Faculty in the Sciences

While there are a number of challenges for female faculty in the sciences, there are several unique issues for those at the senior level. When this group started their academic careers, they were often the first women in their departments, and sometimes the first woman in the science division. This meant that they were frequently asked to serve on college/university committees to provide gender balance. They were often sought after more frequently than other female faculty because they could represent both the perspectives of women and of scientists. That pattern of service has continued throughout their careers. And now, even if they are no longer the only female in their department, they often feel a responsibility to protect their junior colleagues in ways that they were not protected. So, the service load continues.

Service obligations for senior faculty are not limited to their own institutions. As they become more visible, they are often asked to provide service on panel reviews, to serve on advisory boards, and to act as consultants. Upon reflection, one of the co-authors realized that she has served on twenty departmental/divisional reviews with additional ones scheduled for the fall.

Table I. Examples of productive and nonproductive bias avoidance behaviors (10)

Productive	Nonproductive
Delay of partnering/marriage	Hiding caregiving responsibilities
Delay of childrearing	Shirking caregiving responsibilities
Limiting the number of children	Opting out of the tenure track

Again, because the pool of senior female science faculty is small, that group gets tapped more frequently than do male colleagues. Senior female professors may also feel a stronger sense of responsibility to participate in these activities because they feel that some of the activities, such as departmental reviews and advisory boards, offer an opportunity to mentor female faculty in departments in which they may have no female colleagues.

The importance of mentoring cannot be overstated, particularly for women in the sciences who are under-represented in the professoriate. The collegial network provides important information for professional success such as guidance in seeking research funding and advice for establishing a research program. That network can also serve as an important source of referrals for service on advisory boards, and female faculty who are not well mentored can be at a disadvantage (12). The mentoring by female colleagues has been shown to provide different support for junior female colleagues than that of male colleagues. That is, women with effective female mentors feel more empowered and influential in their departments. On the other hand, male mentors are perceived as more important in supporting objective goals such as increased pay and promotion (13). Again, because the group of senior female faculty in the sciences is small, the expectation of mentoring activity falls more heavily on their shoulders.

Finally, because female scientists often put off starting a family, they are more likely to be part of the "sandwiched-generation caregivers", with responsibility for caring for both children and aging parents. This responsibility falls more heavily on the shoulders of women than men, although men are becoming responsible for a larger share of this burden than in the past (11, 14). One study, which focused on couples, showed elevated levels of depression for both caregiving husbands and wives, with greater levels for wives (14). Female caregivers also had higher levels of absenteeism than did their husbands. The study reported that the mutual support offered by the couple was important in dealing with the stresses of caregiving. While single faculty may not be faced with the challenges of dual-care issues for children and parents, the stress of caring for aging parents without the support system of marriage can present particular challenges. These challenges are more acute for female science faculty as a group because they are more likely to be single. Furthermore, single faculty may be asked to do more at work because they are perceived to have fewer family obligations.

Seeking Balance

The critically acclaimed movie "Citizen Kane" opens with the death of newspaper magnate Charles Foster Kane, whose last word is "Rosebud..." The rest of the film follows a reporter trying to uncover the meaning behind Kane's dying word, tracing his rise from poverty to become one of America's most influential men, only to die embittered and alone. Eventually, it is revealed that Rosebud was the name of Kane's childhood sled, suggesting that on his deathbed, his regrets are about his personal life, rather than his professional life. An important lesson from this classic film is to reflect on how you want to be remembered at the end of your life and adjust your priorities appropriately.

In an ideal world, we would have time for everything that matters to us, but in reality, we each need to find the unique balance between our personal and professional lives to meet our own needs. In the long run, such balance is beneficial not only to the individual but also to her colleagues, students, and family members. That is, skills acquired in one arena often make you more effective in another. As an example, many abilities honed through motherhood are extremely helpful in an academic setting, including functioning when exhausted, multi-tasking, learning to be more flexible, and sometimes settling for "good enough". In other words, "If you can manage enthusiasm for *Candyland*, you can manage alertness for the most petrifying committee meeting about copy machines" (*15*). Moreover, children of scientists benefit by their mothers being able to help with most homework, being good role models in a largely male-dominated field, and having skills that can be useful in schools and with extracurricular activities.

Several tips are presented below that may assist those wishing to improve the balance in their lives. First, it could be helpful to reflect on just how much chaos you can comfortably tolerate (*16*). Are you the type who begins projects well ahead of deadlines, or do you work better under time pressure? Do you like to have several projects going on at once, or do you prefer to complete one task before beginning another? Gaining insight into one's tolerance for chaos can be useful to plan your time in your own optimal way, thereby relieving stress. High-energy individuals who thrive when they are busy may be bored when there is too little chaos in their lives, whereas those who don't like surprises may have a low tolerance for chaos. One of the authors of this paper was initially surprised to find that she had a very low "chaos coefficient" but then realized that this preference for control explained a significant amount of stress in her life. (Predictably, she began writing this paper three months before the deadline.)

Reflection on how you are spending your time versus how you would like to be spending your time can also be a useful exercise (*16*). For example, you might discover that you spend more time on housework than you had thought, thereby curtailing discretionary time that could be spent with your family or on a hobby. A possible solution would be to hire a cleaner. One member of our alliance was surprised to find that reading the newspaper was actually one of her hobbies, a fact that she would not have discovered without accounting for time spent during a typical day.

Several useful tips for better integrating the personal and professional life are outlined below, with our own commentary relevant to scientists at liberal arts

Table II. Examples of bias avoidance behaviors exhibited more frequently by women (*10*)

Behavior
Stayed single
Had fewer children than I wanted
Did not ask for a reduced teaching load when I needed it for family reasons because of possible repercussions
Delayed second child until after tenure
Did not ask to stop the tenure clock although it would have helped me
Missed some of my children's important events because I didn't want to appear uncommitted to work
Came back to worker sooner than I would have liked after the birth of a new child

colleges. Many are found within Harvey and Herrild's excellent self-help book (*16*), while others have been learned through our own combined fifty-plus years of experience.

Prioritize

During the semesters, teaching must come first. While class preparation can easily expand to fill all available time, grading and meeting with students are also essential tasks. Maintaining a stable teaching load by teaching the same courses as much as possible can help increase efficiency and decrease preparation time. Try not to let course preparation consume entire summers and breaks, but instead dedicate a specific amount of time to organize your courses (*17*). Summers and breaks provide the opportunity to mentally focus on research projects in larger blocks of time than during the semesters. Research projects should therefore be given top priority when you are not teaching.

Get Organized

Maintain a long-term calendar in which you record class times, meetings, and important deadlines (*17*). Try to leave time before each class to focus. Make sure that you schedule personal time on this calendar, including time for fitness sessions, lunch, medical appointments, and whatever else your own particular needs include. That way, when someone tries to schedule you for those times, you can legitimately tell him or her that you have something on your calendar. From your calendar, create daily and weekly to-do lists. When you complete a task, you get the satisfaction of crossing it off the list. Try to spend the majority of your workday on the critical, although every academic knows that entire days can be spent putting out unforeseen fires. As much as possible, deal with routine paperwork immediately; if you have to return to a document, then you've wasted time. Part of good organization includes cleaning up your office periodically so that you spend less time finding things.

173

Turn Off Your Email Alarm

Most of us have automatic email checking programs that alert us when mail has arrived. While diversions are often welcome during unpleasant tasks, it is better to turn off such a program in order to focus completely on an important job like grading. It is still necessary to check email several times daily, however. We certainly expect to be able to reach students and colleagues and hear back from them in a timely manner, so we must do the same ourselves. When we do have a few minutes to devote to email, it is best to deal with any simple requests immediately, before they scroll off the screen. Remembering an email request that was not fulfilled (for example, for a letter of recommendation already written) causes stress and wastes time searching for the original request.

Spend the Minimal Time Necessary on Routine Tasks

Some tasks are multi-hour jobs, whereas others are 10-minute jobs. Learn to recognize the difference between them and apply your efforts accordingly. Don't spend all day on a 10-minute job! Use a template whenever possible, re-using boilerplate material for standard, low-priority documents.

Avoid Unnecessary Meetings and Lead Effective Ones

While it is not only impossible to avoid meetings completely, but also undesirable in an environment where faculty input is so critical, don't feel the need to attend every campus forum. Again, distractions from unpleasant tasks can be welcome, but the job will still be waiting for you when you return. If you are the person leading a meeting, you can save everyone time by being an effective chairperson. Have an agenda and a timeline, start and end on time, don't recap information for latecomers (which rewards them for bad behavior), keep the meeting brief and to the point, and always end with action steps. In some cases, it may be possible to actually have a "standing" meeting, where business is decided upon without people sitting down and getting comfortable for the long haul (*18*).

Learn the Necessary Skills for Your Job

Having to rely on help from others, such as informational technology or the audiovisual department, often wastes time. Learning how to create and modify your own web pages or how to troubleshoot classroom computer equipment is likely to be more efficient and less stressful in the long run.

Think before You Say "Yes"

Although service is a factor in tenure and promotion, there are only 24 hours in a day (*17*). If you take on a major new responsibility, such as chairing a department or program, you will need to drop some other commitment. Say "yes" strategically, filling roles that will benefit from your special skills. A good rule of thumb is to wait at least a day before agreeing to a new responsibility. Don't feel that you need to justify turning down a new commitment: remember that "No" *is* a complete sentence (*18*).

Get Good Help

Many women perceive asking for help as a sign of weakness. Particularly for those with low chaos coefficients, relinquishing control can be stressful. However, appropriately delegating tasks and sharing responsibilities, both at work and at home, can significantly lighten your load. In terms of getting *good* help, most of us have little control over who our coworkers are, but we can select our own research students. Seek out research students who have shown their ability to work independently, perhaps by asking for referrals from laboratory instructors in the introductory courses. Time invested in training a good student early in their career will ultimately yield a higher payoff than a similar student later in their career. Furthermore, hiring reliable help for household tasks that yield little personal satisfaction can free up time for more important work. Indeed, outsourcing core housework such as cleaning is characteristic of highly productive science faculty (*11*).

Give Up Perfectionism

One of the authors remembers spending several hours baking and decorating loon cookies for her daughter's preschool class on her birthday. While these cookies were indeed lovely, the children would undoubtedly have enjoyed a quicker recipe just as much! Perfectionists may secretly be afraid to submit grant proposals or manuscripts for fear of rejection, but you can't win the lottery if you don't buy a ticket. Learn to recognize when additional time spent fine-tuning is no longer a good investment.

Be Creative in Your Assignments

Professors can be their own worst enemies. Every assignment that a student completes must be graded. By thinking creatively about learning goals, it may be possible to design exams, homework assignments, and laboratory reports that are less time consuming to grade. For example, lab reports that are written as short "communications" not only require students to develop concise scientific writing skills but also require significantly less time to grade than a lab notebook. Using a mixed format for exams, including multiple choice and short answer problems in addition to longer problems, prepares students for standardized tests such as the MCAT and GRE while also decreasing grading time. Instead of grading entire problem sets, consider collecting only one or two problems at random. Using a grading scale of check-plus-to-check-minus instead of actual points on problem sets can also save grading time. Students often don't read written comments on their work and are likely to be just as satisfied with being referred to a written answer key as being given a detailed explanation of why their answer is incorrect.

Leave the "One More Thing" until Tomorrow

How many times have you been ready to leave for an appointment but then realized that you had 5 or 10 minutes to spare? While that seems like long enough to send just one more email or look up a reference, you often end up rushing off to your appointment and being late anyway. Don't try to squeeze in one more task on your way out. It is much less stressful, and more professional, to be on time than to be late. Besides, it often takes at least 5 minutes to pack up and get out the door anyway!

Leave Time for Yourself

Most of us feel that we don't have time to schedule exercise, doctor's appointments, or even haircuts during the semester. However, in order to perform our best, we need to feel our best. When was the last time you saw a college administrator who needed a haircut? Put personal time on your calendar! You deserve it.

Conclusions

Carving personal time out of a busy workday is a constant battle for science faculty, but senior women face even greater challenges to do so. Nonetheless, the creative freedom of an academic career makes it possible to set one's own priorities more so than in many other professions. Overcoming the guilt of dividing your energy between multiple life roles is an excellent first step in achieving balance. Ultimately, the most sustainable lifestyles are those that set sufficient boundaries to incorporate your own personal needs into the equation.

Acknowledgments

We thank the other members of our alliance, Janis Lochner of Lewis and Clark College, Joanne Stewart of Hope College, and Melissa Strait of Alma College, without whom this work would not have been possible. We also thank Cathy Bevier of Colby College for helpful suggestions. We acknowledge funding from the National Science Foundation ADVANCE Partnerships for Adaptation, Implementation, and Dissemination (PAID) Awards: NSF-HRD-061840, 0619027, 0619052, and 0619150.

References

1. Carroll, J. Gallup: Time Pressures, Stress Common for Americans. http://www.gallup.com/poll/103456/time-pressures-stress-common-americans.aspx#1 (accessed June 2010).
2. Salholz, E. Too Late for Prince Charming? *Newsweek*, June 2, 1986, p 54–61.
3. Bloom, D. E.; Bennett, N. G. *Marriage Patterns in the United States*; NBER Working Paper Series, No. W1701; National Bureau of Economic Research, Inc.: Cambridge, MA, 1985.

4. Brooks, A. Relationships: When Studies Mislead. *New York Times*, Dec 29, 1986.
5. McGinn, D. Marriage by the Numbers. *Newsweek*, June 5, 2006.
6. Landau, S. Tenure Track, Mommy Track. *Association for Women in Mathematics Newsletter*, May–June 1991.
7. O'Reilly, A. In *Parenting and Professing: Balancing Family Work with an Academic Career*; Bassett, R. H., Ed.; Vanderbilt University Press: Nashville, TN, 2005; pp xiii–xvii.
8. Cohen, H. The Baby Bias. *New York Times*, August 4, 2002.
9. Correll, S. J.; Benard, S.; Paik, I. *Am. J. Sociol.* **2007**, *112*, 1297–1338.
10. Drago, R.; Colbeck, C. L.; Stauffer, K. D.; Pirretti, A.; Burkum, K.; Fazioli, J.; Lazzaro, G.; Habasevich, T. *Am. Behav. Sci.* **2006**, *49*, 1222–1247.
11. Schiebinger, L.; Gilmartin, S. K. *Academe* **2010**, *96*, 39–44.
12. Rosser, S. V.; Tayor, M. Z. *Academe* **2009**, *95*, 7–10.
13. Settles, I. H.; Cortina, L. M.; Stewart, A. J.; Malley, J. *Psychol. Women Q.* **2007**, *31*, 270–281.
14. Hammer, L. B.; Neal, M. B. *Psychol. Manager J.* **2008**, *11*, 93–112.
15. Holloway, R. In *Parenting and Professing: Balancing Family Work with an Academic Career*; Bassett, R. H., Ed.; Vanderbilt University Press: Nashville, TN, 2005; pp 93–101.
16. Harvey, C. S.; Herrild, B. E. *Comfortable Chaos*; Self-Counsel Press: Bellingham, WA, 2005.
17. Karukstis, K. Time and Stress Management for Faculty in the Liberal Arts College Environment. NSF-ADVANCE: Collaborative Research for Horizontal Mentoring Alliances. http://www.hmc.edu/files/chemistry/KKKTimeManagement082609-1.pdf (accessed June 2010).
18. NSF ADVANCE Summit on the Advancement of Senior Women Scientists at Liberal Arts Colleges, Washington, DC, June 2–4, 2010.

Chapter 13

Enhancing Your Professional Presence

Julie T. Millard*

Department of Chemistry, Colby College, Waterville ME 04901
*jtmillar@colby.edu

Leadership roles in science departments have traditionally been occupied by male faculty members, and unwritten measures of performance may limit the professional influence of women not privy to a support network. This paper addresses some of the factors that impact how women are perceived in an academic setting and presents some strategies for enhancing one's professional presence.

Establish Your Professional Presence

Back in high school, I was a state finalist in a public speaking competition. After I finished second by one point, a woman from the audience approached me and identified herself as a speechwriter for our senior senator. She told me that I would have won the competition if I had been wearing a skirt instead of pants. This was my first lesson that you may be judged by more than what you say and how you say it. From that moment on, I paid more attention to my appearance, especially when the stakes were relatively high.

For most women scientists, the term "professional presence" may seem more relevant to those in business than in academia. Indeed, the rules that govern professional appearance appear to differ for women in academia relative to other workplaces, where makeup use is often associated with success (1). However, as academic women, we are constantly being judged by our students, colleagues, administrators, and peer reviewers, who often use criteria unrelated to our knowledge or accomplishments. Simply being aware of some of the factors by which we are evaluated can be helpful to make adjustments to enhance our professional presence.

© 2010 American Chemical Society

First Impressions Matter

A familiar saying is that you never get a second chance to make a first impression. An important opportunity to make a good impression is the first day of class, when students often make quick judgments about an instructor's competence (2). One study that surveyed college students after the first class found that their initial assessments were based primarily on "communicative competence", including the instructor's overall speaking ability, verbal and nonverbal communication skills, ability to adapt material to the students' knowledge level, level of clarity, organizational skills, and ability to generate interest (3). As professors, we would certainly expect students to value these traits. However, perhaps more surprisingly, another study found that 10-second silent video clips are good predictors of end-of-semester evaluations (4). Teaching evaluations were found to correlate both positively and negatively with many types of nonverbal behavior (Table I). Physical attractiveness was also found to influence student ratings somewhat (4). Other studies have also found that factors not directly related to learning may influence student ratings, including warmth and friendliness, particularly for female faculty (5).

Effective Nonverbal Communication

"Act out being alive, like a play. And after a while, a long while, it will be true."

-John Steinbeck, *East of Eden*

Many of us are largely unaware of the subtle signals communicated by our body language and other nonverbal cues. However, simply acting more optimistic or enthusiastic can influence others to perceive you that way. In a dramatic illustration of the power of nonverbal communication, a psychology professor at Cornell attempted to teach the identical course in the fall and spring semesters, with one exception (6). The second time, he adopted a more "enthusiastic" style, varying his vocal pitch and using more hand gestures. Rather to his surprise, student ratings of his course improved significantly, with higher scores in the spring semester for his level of knowledge, organization, accessibility, fairness, and even the quality of the textbook, which was the same. Students also reported increased learning, although test performances and final grades were virtually identical in both courses.

In addition to being enthusiastic, your nonverbal communication can be enhanced by making good eye contact, having good posture, and displaying energy and confidence (7). A positive, can-do attitude will also reflect positively on you: remember that nobody likes a complainer! Indeed, perceived optimism had the highest correlation with college teacher effectiveness ratings (4).

A professor's classroom attire can also influence student perceptions. Faculty members who dress formally are generally perceived as being more organized, knowledgeable, and competent, whereas those who dress casually are viewed as friendlier, more approachable, and more willing to listen to student opinions (8–10). It would seem strategic for younger female professors to dress "professionally" in order to increase their perceived competence, particularly in male-dominated fields such as the sciences. On the other hand, older female professors who are found unapproachable might try more casual attire if their competence is no longer in question. Even though academics have a reputation for dressing poorly relative to other professionals (*11*), it may be worth the effort to spruce up your personal appearance and convey a more professional look (see (*12*) and (*13*) for some excellent tips). For example, every professional woman's wardrobe should include a tailored black suit and plain black pumps (*13*). Again, academics are not generally held to as high standards for good fashion: although the highest-paid female executives frequently wear the highest heels, female academic rarely wear high heels (*11*), particularly not in a laboratory setting.

A largely neglected arena through which professors make an impression is their office doors. One study focused on the occupational identities displayed through academic door displays, finding relationships between these expressions of self and social status (*14*). Hallway bulletin boards are also an important venue to influence how an individual, a department, or a program is perceived. Particularly for high-traffic areas, these displays can influence whether a prospective student chooses to attend an institution, selects a major program of study, or elects to work in a particular research laboratory. Look critically at such displays in your hallways to determine if they are consistent with the mission of your laboratory, department, or program. Do they reflect the identity by which you wish others to perceive you?

Self Promotion

As observed by Peggy Klaus (*15*), "Promoting ourselves is something we are not taught to do. Even today, we still tell children 'Don't talk about yourself, people won't like you.' So ingrained are the myths about self-promotion, so repelled are we by obnoxious braggers, many people simply avoid talking about themselves." Instead, many of us believe that we will receive recognition for the good work that we do. Sometimes this is true, and certainly one of the positive actions we can take for a colleague is to tell the world about her latest accomplishment. However, we can't simply assume that our work will always speak for itself, and sometimes it is necessary to let the chair or the dean know about the acceptance of a major paper, the funding of a grant proposal, or an invitation to speak at a conference. Men are often better than women at self-promotion (*15*). A former chair of my department told me that he was surprised to learn from my annual report that I had more papers than one of my male contemporaries. My colleague had promoted his own work so effectively that everyone assumed that he had published far more than he actually had.

Part of getting the word out about your accomplishments can be through a network. For example, some of the women on my campus recently initiated a program called *Women in the Spotlight*, a monthly celebration of the accomplishments of women at our institution (*16*). In addition to raising the public profile of the women on campus, the organizers also hoped to foster dialogue and excitement about women's contributions to intellectual life and community. Venues such as these are an important first step in spreading the word about the work that women faculty are doing.

Table I. Examples of nonverbal behavior that correlate with good teaching evaluations (4)

Behaviors with positive correlation	Behaviors with negative correlation
Projecting optimism	Frowning
Projecting confidence	Fidgeting
Projecting enthusiasm	Sitting

Establishing contacts in the media can also help you to become known in the larger community. Alert them to significant accomplishments that might be of interest to their readers. Since you never know when you will meet someone influential, be prepared with an "elevator speech" that sums up your work in clear and concise language. The general idea behind an "elevator speech" is that you should be able to promote yourself to someone you encounter during an average elevator ride (about a minute in length). You might want to prepare two versions of your speech, one for the layperson and another for someone with prior knowledge of your field.

Also try to develop your network of contacts at professional meetings. Individuals working in similar fields can be excellent reviewers of your work, and if they have a positive image of you from a brief encounter, then so much the better! Although business cards may not be as popular in academia as in other professions, your card contributes to the first impression you make. It should be up-to-date, and you should always have a supply with you.

Maintain a Strong Virtual Presence

In these times of electronic networking, many first impressions are made in the absence of face-to-face interactions. As an example, a female colleague at a liberal arts college received an email from a graduate student requesting that she send him a plasmid created in her laboratory for use in a particular experiment. She responded that she had intended to do a similar experiment herself but would be amenable to collaboration. Shortly thereafter, she received an email from the major professor, who had unintentionally included her in his response to his graduate student. The professor had checked out my colleague's laboratory webpage and said, "She's good. Let's play ball." The time she had invested in her laboratory webpage paid off.

Web Presence

Make sure that you not only have a webpage, but also that it enhances your professional presence. If someone runs your name through his or her search engine, what pops up? The information on your webpage should be no more than a year old, so make sure that you set aside time each year to update your recent accomplishments. The summer is an excellent time to do this. Additionally, if you have a Facebook page, be prudent about what kind of information you post, keeping in mind the worldwide presence of such social networking sites. You never know who might take a look at your page.

Telephone Presence

Always keep in mind that your first impression could also be made over the telephone. Evaluate before you answer the phone; if you don't really have time to talk, then let it go to voicemail rather than being short with the person calling. Telephones with caller identification are becoming more common and are useful for screening your calls. (Do you really have the energy to talk to the parent who has already called you several times about her son who is failing your course?) If you have important business to discuss over the telephone, make an appointment so that you won't be rushed. Put a sign on your door saying "important phone call in progress" to avoid being disturbed.

While hiring student workers, a colleague once called the next person on her list, only to reach a voice mail greeting filled with profanity. Instead of offering the student a job, she left a message saying that she had crossed him off the list of possible student workers. Clearly, faculty members will not have such offensive greetings, but we should keep our outgoing messages professional, short, and friendly. If you will be out of the office and not checking messages daily, you should update your greeting to reflect that. However, make sure that you promptly change your greeting when you return, or you will lose credibility.

When you leave voicemail messages for other people, keep them short and to the point. Briefly identify yourself, give your contact information, and state the purpose of your call so that there is incentive to get back to you. Never leave a harsh voice mail message; you may regret it later.

Email presence

Every email that we send makes an impression on the receiver (7). With the rise in the popularity of electronic communication methods such as text messaging and tweeting has arrived the emergence of a new language of abbreviations that may soon hamper our ability to communicate with our students. However, faculty who are up to date with chat abbreviations should avoid using them in professional contexts. Construct your email messages similarly to written notes, using complete sentences, correct spelling, and proper punctuation.

People often write things in an email that they wouldn't normally say face-to-face, a phenomenon known as the "online disinhibition effect" (17). Be careful of any email that you send, always remembering that your message could be forwarded, even to the person you just spent a paragraph complaining about. Never send an email when you are angry, and, like any written document, proofread before sending. Give your email a brief but informative subject heading that will compel the recipient to actually read it.

Be sure that you are actually sending the email to the person that you think you are. Most of us have had at least one embarrassing incident when we sent an email to the wrong person. "Reply all" can be a dangerous tool. Also keep in mind that copying the dean is a power play that suggests a lack of trust and may annoy the person with whom you are corresponding, unless your email is a positive one (7). Copying the dean can also be a useful move when you feel that you must go over someone's head in order to get the desired outcome. If you are sending a warning or disciplinary email to a student, make sure that you copy the chair, the student's advisor, and yourself so that you have a written record of the correspondence.

Finally, make sure that you check your email regularly and respond in a timely fashion. People who don't respond to emails are often perceived as unprofessional and unreliable. As with your voicemail greeting, if you will be unavailable for more than a few days, set up an autoreply message stating when you expect to be able to respond.

Demonstrate Good Leadership Skills

Once you are tenured, you can expect your service load to increase, even though it has probably already been higher than that of your average male colleague (18). The nature of your service will also change, with increased roles in leadership. Indeed, opportunities for leadership may be more common for women scientists at liberal arts colleges than at research universities. In 2002, women held only 4.6% (26 of 566) of the department chair positions at R1 university math and physical science departments (19). However, at a recent conference of senior women scientists at liberal arts colleges, virtually every woman had served as chair of her department (20). Women tend to be different types of leaders than men, often working towards a consensus rather than using their authority to make unilateral decisions (21). Women also tend to be less confrontational, preferring to respond with written comments and criticisms rather than with face-to-face conflict resolutions. Furthermore, female chairs may be perceived as more approachable by those seeking to secure job benefits for themselves (22). Serving as chair can lead to increased influence in one's institution, particularly for women, who often have less power in academic departments than men (23). Senior women scientists generally have the skills needed for leadership, having survived a highly competitive, male-dominated system (22). However, because of fewer opportunities for networking, they may be less confident in leadership positions.

Manage Effective Meetings

One of the first skills to develop in a leadership position is the ability to run an effective meeting. First of all, consider whether there is actually a need to meet. Just because your department has always had weekly meetings doesn't mean that you actually need to meet that frequently. Perhaps every other week or even once a month is sufficient. Decide on the meeting frequency before the start of each semester so that everyone can put the meetings on their calendar and minimize conflicts. A few days before a meeting, send an email agenda to all participants, giving them an opportunity to add any new business. This email also serves as a reminder about the meeting.

Model conduct that you expect from your colleagues. For example, make sure that you both arrive on time and start on time. Setting clear goals and focusing on achieving them facilitate ending the meeting on time. Circulate necessary materials before the meeting, and make it clear that everyone is expected to review this information before the meeting. Be prepared to ask your colleagues to share responsibility: just because you are chair doesn't mean that you have to do all the work. Keeping a written record of department jobs and who holds them can be useful in identifying those who have lighter loads and could therefore take on new tasks. At the end of meetings, review actions to be taken and the person who will take responsibility for implementation.

Beware of spontaneous hallway meetings, which may be expedient but could lead to disenfranchisement for department members who feel like their voices were not heard. However, speaking informally with key department members about issues ahead of time can be strategic to build consensus on important issues.

Engage in the Workplace

Part of good leadership is being positive about the department and institution you represent. If you show team spirit, then others will want to join the team. When you meet with prospective students and job applicants, be enthusiastic, rather than on dwelling on any negative aspects. On any team, the success of an individual reflects positively on the group, so give credit where credit is due. Post the accomplishments of faculty members and students on the department webpages and/or bulletin boards. This will encourage others to do something worthy enough to be mentioned themselves.

Even though balancing an academic position and a personal life can be challenging (18), you should make an effort to participate in departmental social events when possible, including parties, outings, and retreats. Your presence may be of higher value when students are involved. Participating in college functions such as dinners with the trustees and administration may also raise your institutional profile, and thus increase your influence.

Lead by example, showing your students that you are actively engaged in teaching, scholarship, and service so that they see the value in these activities. Try not to hide your personal responsibilities and life choices; for example, bring family members to work-related social functions. A more family-friendly workplace climate can be created through the actions of those who explicitly acknowledge their own family commitments and/or provide support for the personal lives of their colleagues (24).

Engage in the Community

Take the opportunity to contribute to the community in your own special way by volunteering to give talks in your area of specialty or serve in leadership roles. For example, one of my colleagues serves as a HazMat officer for the local fire department, where his skills as a chemist make him highly valued. Many organizations would be happy to have help from a successful woman scientist with an abundance of valuable skills.

Conclusions

As the number of senior women scientists at liberal arts colleges increases, we can expect more females in leadership positions. Being aware of factors that influence how others perceive us can help maximize our impact, both on our own campuses and in the larger community. Attention to professional development can enhance one's visibility, productivity, and leadership, in turn enhancing job satisfaction (25). Women scientists in leadership roles can also promote practices that result in more female-friendly departments, leading to a more rewarding work environment for all (24, 26).

Acknowledgments

I thank the other members of my alliance, Janis Lochner of Lewis and Clark College, Nancy Mills of Trinity University, Joanne Stewart of Hope College, and Melissa Strait of Alma College, without whom this paper would not have been possible. I also thank Paul Greenwood and Cathy Bevier of Colby College for helpful suggestions and acknowledge funding from the National Science Foundation ADVANCE Partnerships for Adaptation, Implementation, and Dissemination (PAID) Awards: NSF-HRD-061840, 0619027, 0619052, and 0619150.

References

1. Dellinger, K.; Williams, C. L. *Gender Soc.* **1997**, *11*, 151–177.
2. Brooks, D. M. *Theory Pract.* **1985**, *24*, 63–70.
3. Hayward, P. You Never Get a Second Chance To Make a First Impression. Annual Meeting of the International Communication Association, Dresden International Congress Centre, Dresden, Germany, 2009. http://www.allacademic.com/meta/p74696_index.html (accessed June 2010).
4. Ambady, N.; Rosenthal, R. *J. Pers. Soc. Psychol.* **1993**, *64*, 431–441.
5. Kierstead, D.; D'Agostino, P.; Dill, H. *J. Educ. Psychol.* **1988**, *80*, 342–344.
6. Williams, W.; Ceci, S. *Change* **1997**, *29*, 12–23.
7. Bixler, S.; Scherrer Dugan, L. *5 Steps to Professional Presence*; Adams Media Corporation: Avon, MA, 2001.
8. Roach, K. D. *Commun. Q.* **1997**, *45*, 125–141.
9. Dowling, W. A. *Coll. Teach. Methods Styles J.* **2008**, *4*, 1–11.
10. Lavin, A. M.; Carr, D. L.; Davies, T. L. *Res. Higher Educ.* **2009**, *4*, 1–15.
11. Steele, V. The F-Word. http://www.wiu.edu/users/mfbhl/180/steele.htm (accessed June 2010).
12. Gross, K. J.; Stone, J. *Chic Simple Dress Smart for Women: Wardrobes That Win in the Workplace*; Grand Central Publishing: New York, 2002.
13. Finney, K. *How to Be a Budget Fashionista: The Ultimate Guide to Looking Fabulous for Less*; Ballantine Books: New York, 2006.
14. Miller, J.; Behringer, A.; Anderson, E. K.; Bovard, S. E.; Brimeyer, T.; Grantham, R.; Li, Y.; Osborne, L.; Reyes, Y. *Soc. Sci. J.* **2005**, *42*, 459–467.

15. Klaus, P. B. *Brag! The Art of Tooting Your Own Horn without Blowing It*; Warner Business Books: New York, 2003.
16. Collins, S. Women in the Spotlight. *Colby Magazine*, Fall 2009, p 8.
17. Suler, J. *CyberPsych. Behav.* **2004**, *7*, 321–326.
18. Millard, J. T.; Mills, N. S. Integrating Work and a Personal Life: Aspects of Time and Stress Management for Senior Women Science Faculty. In *Mentoring Strategies To Facilitate the Advancement of Women Faculty*; Karukstis, K., Gourley, B. L., Rossi, M., Wright, L. L., Eds.; ACS Symposium Series 1057; American Chemical Society: Washington, DC, 2010.
19. Shauman, K. A. Executive Summary: Women, Work and the Academy. Barnard Center for Research on Women. http://www.barnard.edu/bcrw/womenandwork/shauman.htm (accessed June 2010).
20. NSF ADVANCE Summit on the Advancement of Senior Women Scientists at Liberal Arts Colleges, Washington, DC, June 2–4, 2010.
21. Alimo-Metcalfe, B. *Women Manage. Rev.* **1995**, *10*, 3–8.
22. Yentsch, C. M.; Sindermann, C. J. *The Woman Scientist*; Plenum Press: New York, 1992; pp 125–144.
23. Simeone, A. *Academic Women: Working towards Equality*; Bergin and Garvey: South Hadley, MA, 1987; pp 87–90.
24. Drago, R.; Colbeck, C. L.; Stauffer, K. D.; Pirretti, A.; Burkum, K.; Fazioli, J.; Lazzaro, G.; Habasevich, T. *Academe* **2005**, *91*. http://www.aaup.org/AAUP/pubsres/academe/2005/SO/Feat/drag.htm (accessed June 2010).
25. Karukstis, K. K. Faculty Engagement and Career Satisfaction at Liberal Arts Colleges. NSF ADVANCE Summit on the Advancement of Senior Women Scientists at Liberal Arts Colleges, Washington, DC, June 2–4, 2010.
26. Gourley, B. L. Leadership Support for Women Faculty Members in Science, Technology, Engineering and Mathematics (STEM) Disciplines at Liberal Arts Colleges (LAC): Perspectives on Practices, Policies and Infrastructure Related to the Position of Department Chair. NSF ADVANCE Summit on the Advancement of Senior Women Scientists at Liberal Arts Colleges, Washington, DC, June 2–4, 2010.

Indexes

Author Index

Subject Index